The Machiavellian Engineer

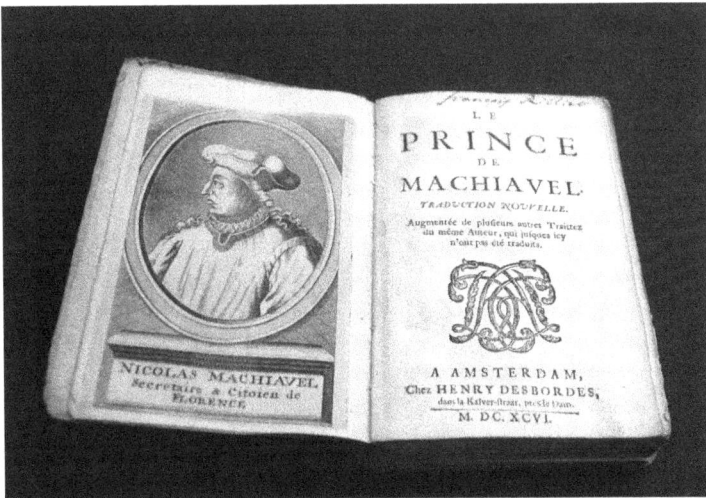

Max Cavelli

The Machiavellian Engineer

http://maxcavelli.com/

© 2024 Shivkumar Iyer
Print ISBN ISBN 978-1-960405-29-6
ebook ISBN 978-1-960405-30-2

Cover Design by Guy D. Corp
www.GrafixCorp.com

STAIRWAY PRESS—APACHE JUNCTION

STAIRWAY≡PRESS

www.StairwayPress.com
1000 West Apache Trail
Suite 126
Apache Junction, AZ 85120 USA

STAIRWAY PRESS · APACHE JUNCTION, AZ

The Relentless Power of Reason

Publisher's Note

TAKE A LOOK around and notice things that are good in your life. When I get up in the morning, I see things like my clock radio, electric toothbrush and razor, the shower head and the hot water coming out of it. While getting ready for work, I will often be listening to a podcast on my cell phone. I have electric lights and air conditioning (I live in the Sonoran Desert, so AC is a key part of my comfort). In my kitchen, I have a coffeemaker, toaster, refrigerator, gas stove and dishwasher. The list goes on and on.

What do all these things have in common?

They are all engineered.

Somewhere in the world, mechanical engineers, electrical engineers, software engineers, circuit board designers and other technical people focused on solving problems for my comfort and safety. Civil engineers played a role in getting water and other utilities to my house. Sanitary engineers designed the underground pipes and pumps to take care of my sewage.

I have all these useful things because of engineering.

These things were not created by Hollywood (or Bollywood) actors. They were not created by football players. They were not created by politicians.

If you enjoy comfort in your life, overwhelmingly it's due to engineering. Am I wrong? Think about it.

Other people play a role, but they are not as important. People must build and transport the stuff and projects must be funded and managed. Those tasks have value. No disrespect intended. But, without the brainpower and skill of engineering, your life would be harder. You would die at a younger age. More of your children would not survive to adulthood.

What are common traits of engineers?

They are smart.

They are motivated by challenges.

They are educated.

They are special.

So, why is it that you have favorite actors and athletes? Why do you know the names of famous politicians…politicians who can only hinder and destroy, not create?

Engineers can have very good lives. I certainly have nothing to complain about. Fifty years of being an engineer was very good for me.

However, as Max Cavelli documents in this book, company bureaucrats and managers ruin what should be a wonderful and respected career. Give engineers technical problems to solve and a budget for getting things done and they are happy. Bog them down with stultifying overhead and nonsensical busywork and they are unhappy.

Why for one minute should the people most responsible

for the good things in your life be unhappy? We should be feted and treated like royalty.

But, often, we are not.

I truly wish Max's book was unnecessary. I would love to say, "Max, you have a great life and the world appreciates everything you do, so don't be so cynical and downbeat."

But that's not the world as I experienced it. The company I work for now values engineering and I have a great boss who lets his engineers be engineers, but that's unusual.

I would love to tell you this book is not needed. I would love to tell you that manipulation and subterfuge will not be a part of your worklife.

That's not reality.

It would be great if taking the high road, applying your intellect, and working hard was sufficient for building a satisfying and productive career. You'd get a lot done and the world would be a better place.

The least you need to know is what Machiavellian strategies might be used against you. You should be aware so you can see what is happening and make an intelligent plan to do something about it. That's a minimum takeaway from this book.

Do you want to go further? You can join the game, fight back, and use Machiavellian tactics to win against the dark forces that would make you a slave. Read Max's book. Think about what Machiavelli says. Take pride in how much your profession means to the world and don't let office politics, bullies and second-handers slow you down.

Engineers are the real superheroes.

That reminds me. I need to stop at the cleaners and pick up my cape.

—Ken Coffman
Publisher, Stairway Press

Introduction

WHY WOULD ANYONE bother to write another book based on *The Prince* by Nicolo Machiavelli? Hasn't this topic been researched enough for other books?

These thoughts might pop into anyone's mind when they read the title of this book. Except, this is not a book for leaders or for those who want to be leaders.

This book is for those with no leadership aspirations. It is for engineers who want to live a peaceful life in their 9-to-5 job. This book is written by someone who lives such a peaceful life in a 9-to-5 engineering job.

This book is intended as an elixir for those who find it agonizing to wake up in the morning, go to work, and later go to bed at night thinking what a waste their life and work is. I figured out a solution to that problem. This book is about carving out a niche of peace and tranquility—not through yoga, meditation, or mindfulness, but through skills, expertise, and creativity.

Machiavelli is usually quoted in the worlds of business or politics. Renowned businessmen, businesswomen and politicians quote Machiavelli's philosophy as a priest would recite the scriptures.

However, not only can Machiavelli be used by those in power but also by those who need to live under those in power, thereby, be able to judge these leaders for their worth and determine how much to invest in them. Quite sadly, it is observed for the most part, we devote the best of our time and energies to the wrong type of leaders—those without potential and who can never adequately compensate us.

It is important to remember that loyalty is a valuable commodity. One should not waste it on the wrong people. Rather, loyalty must be earned, cherished, and rewarded.

How do we decide whether someone is worth our loyalty? The answer lies in interpreting the philosophy of Machiavelli—examining it through another lens, not with the intention of running a business or an organization the way Machiavelli's philosophy usually is, but rather to take back control of our professional lives.

My desire is to resonate with working folk. The only reason for specific reference to engineers is because I am an engineer. I devoted the past twenty-plus years of my life to engineering as a graduate student working under professors in engineering departments or as an engineer in companies.

Just as Machiavelli wrote his book based on his experiences as a statesman, this book is written based on my experiences as an engineer. That said, in no way do I intend to write this book looking down at other professions. It is merely that the examples and perspectives in this book will be in the context of engineering companies, as these are the ones I know.

In today's day of mass social media, one finds a constant stream of advice. This advice is from various sources—recruiters, managers, CEOs, and also a new cohort of career coaches. Many have vast numbers of followers and religiously post pieces of advice either every day or several times a day.

Unfortunately, I am only human and cannot resist being curious about what these wise people might have to say. Some claim to be liberal, some conventional and corporate, some leftist and some reactionary capitalist. The posts are quite often well-crafted and sometimes even professionally edited with the objective of reaching a particular audience using features offered by social media platforms, may it be advertising or boosting a post.

In my observation, many of these Influencers offer advice to young professionals about how they need to make that first step in a giant leap forward. A lot of the advice is obvious and need not really be emphasized—being punctual, being diligent, being enthusiastic, being open to criticism, thinking out of the box, and many others.

On the flip side, anyone with workforce experience can attest to how questionable this advice is.

Over my years as an employee, I was left holding the short end of the stick many times and have seen it happen to others. I have been in extremely toxic workplaces and have had a few good experiences as well. Many have the habit of brushing off bad experiences as simple bad luck and finding ineffectual ways to cope with it, while a few are adaptable and find a way to blend into any organization.

Just as in Machiavelli's turbulent political times, today's corporate world is one filled with hostility and intrigue. Some survive and thrive, while others languish and waste away. It is impossible to design a complete, universal system where everyone can thrive and flourish. Even in the best designed systems, a few will fall through the cracks.

However, when multitudes are failing, one should acknowledge these are not just cracks, but notice the floor is caving in altogether.

When we conclude the nature of work in these modern

times is a mess, the productive option is to look for solutions. We want meaningful work that provides adequate remuneration and additional benefits. To survive and thrive in our world is a skill that must be carefully honed. With awful managers and bad colleagues—we must make do with what we have.

It is difficult to endure working for a manager from hell. Many managers do not fall into that category but are incompetent for many reasons. In the same way Machiavelli discerned between good rulers and bad rulers, it is easy to distinguish a good boss from a bad one. Having made this distinction, the next step is to survive.

Persisting under incompetent or bad bosses is tricky. Unfortunately, most of those whom I know who did so (including myself) learned through the painful experience of being burned at least once.

In such cases, one should change perspective to be a manager of their own lives and careers. Fortunately, this is when Machiavelli's philosophy is apt. Though a lot of it was written almost 500 years back, Machiavelli's philosophy can be applied with careful modification by an employee trying to survive in an organization.

I saw some do so and have done so myself, and, mostly, the results were successful. These experiences led and empowered me to write this book.

A few people I know were aghast that I studied the writings of Machiavelli. They think of him as a teacher of evil. Logically, Machiavelli wrote down what had been practiced at that time by many a ruler. His writings were full of references to historical events—either in the decades before the publication of his works, or in a few cases, going back a few hundred years.

In those days, politics was brutal, unconstrained by law and violence was common, whether in the context of war or via political assassinations.

What makes Machiavelli worth reading are that they contain in-depth analysis of events and was not just a book with instructions. Additionally, his writings compare historical events by describing where one ruler erred while another succeeded.

There are those who see Machiavelli's writings as a guidebook to tyranny and therefore, inapplicable to the modern world. One should take into account that in modern times, though a large part of the world lives in a semblance of democracy with elections taking place regularly and citizens having a right to cast their vote, even devout lovers of democracy will admit it has weaknesses, and with several players determined to weaken it at all costs, democracy favors the powerful.

Therefore, one can examine Machiavelli's writings from a purely political angle to determine how it can be abused, and subsequently, how it can be strengthened. This book is not about politics, but rather about the modern corporate world, and, in this context, Machiavelli's writings are useful.

Most of us need to work to earn a living. Though human beings work under very different conditions, this book will focus on those who work in established businesses or companies, have some form of formal contract with their employers, and provide a certain array of services.

As stated before, being an engineer, I can speak mainly of the engineering world and for those engineers who completed some form of engineering training. Except for a small proportion of engineers who are unionized workers, and therefore, have a union that can negotiate with their employers, most engineers work in companies where they have no say in decisions made by the company.

As an engineer, chances are that one will have to complete assigned tasks without much of say in how those tasks are assigned or how they are to be completed.

Furthermore, the performance of the engineer will be judged by management in a manner decided solely by them. Growth and prospects in the company are decided solely by the management.

It is quite clear that most engineers work in a state where there is little or no democracy with all decisions taken by the management. Though one might argue this is not tyranny, the situation is ripe for all forms of exploitation and abuse. As any engineer who worked in an engineering company can attest, working conditions companies can range from excellent to pitiful.

At one end of the spectrum are progressive companies that value engineers and provide them with excellent facilities to work, and adequately compensate and reward their employees. At the other end of the spectrum are companies managed poorly where employees are harassed and exploited, and where pay is meagre.

Sadly, except for a small proportion of engineers who graduate from the best universities, most engineers can only get low-grade jobs; hoping to eventually climb up the engineering food chain as they gain experience.

Company managers are tasked primarily with maximizing the profits of companies, and towards this end will use every possible technique to cut their costs, while increasing productivity. To achieve this, the common strategy is to pay its employees the minimum necessary to retain them, while extracting the greatest possible work output from them.

In the case of unionized workers, the unions can negotiate with their management to increase the pay and benefits of workers while limiting steep raises in the expectations from them.

However, for those workers who are not part of a union, they have to deal with the management on their own, thereby, leaving them with little choice but to comply with the wishes of the management if they wish to retain their positions.

In addition to having the scales tipped in their favor, company managements have adopted a number of strategies to create competition between its employees to generate an atmosphere of insecurity, where employees constantly have to worry about an axe falling on their head by being made redundant.

Though governments passed legislation safeguarding some of the most basic rights of employees, company managements have evolved increasingly sophisticated techniques to increase pressure on its employees. Profit making companies form the backbone of any country's economy, not only by generating tax revenue, but providing livelihoods to its employees. In most countries, these companies have a fair degree of freedom to operate with only basic safeguards for its workers.

Under these circumstances, it is clear many in the workforce are at the mercy of company management that exists solely to make a profit.

In modern times, many companies adopted ruthless management practices—and many are brazen about their strategies. Increasing pressure on employees results in increased health problems, may they be physical or mental health related. Companies adopting a cut-throat hire and fire policy place their employees under a constant state of insecurity about their jobs.

Widely publicized mass layoffs have employees wondering if they will be the next to face the axe. Though a spirit of competition and accountability is acceptable, the corporate world is a jungle and sets off alarm bells even in proponents of free trade.

With this backdrop, it is evident that Machiavelli's writings are applicable to modern times. Our workplaces are ripe for tyranny. Machiavelli's writings are popular among industry leaders and entrepreneurs who use his techniques to increase profitability.

One could say that—in the absence of legislation and safeguards—company management will become increasingly Machiavellian as his writings are the basis for a leader to attain power. As a first step, an in-depth analysis of Machiavelli's philosophy will help to understand the cut-throat nature of companies that care for little other than profit.

My book offers means by which an engineer can survive in a company whose management is Machiavellian.

Machiavelli's writings are special for several reasons. To begin with, Machiavelli was active in politics during the decades before he wrote his book, being a senior official in the Florentine Republic in the early sixteenth century. Therefore, he had an inside view of the tumultuous years of Italian politics with a keen eye on many of the events that took place during those years.

His writings in *The Prince* feature heavily the key political figures at that time—Cesare Borgia, Pope Julius the Second, the King of France and the King of Spain, along with many of the princes of some of the smaller principalities of Italy.

Therefore, as a political treatise, *The Prince* is detailed and analytical. However, though there were previous political treatises—such as *Artha Shastra* by Chanakya and *The Art of War* by Sun Tzu, Machiavelli stands apart from them as his writings were based on in-depth experiential analysis and followed with tactical instruction.

I read *The Prince* for the first time in the early 2000s when I was a graduate student. Though Machiavelli's writings are difficult to read and comprehend at first, it made a strong impression on me.

What made the greatest impact was the abundance of quotes, many of which are popular and can be found all over the internet. It took many rereads over the years before which I could fully understand *The Prince* and piece together many of the historical events after reading up on them independently.

Since I was merely a graduate student, I had little interest in politics and had read *The Prince* just like one would read a story book. The significance of his writings only became evident when I started working in industry much later—and began to experience the toxicity in the workplace.

In industry, I found the initial few years extremely frustrating. I was unable to cope. My experience was like many of my colleagues. Out of disgust, most of them left the industry—returning to academia. It gradually dawned on me that this loss of talent was not any real and significant loss for management and could be termed a pruning, where management had merely weeded out a few undesirables.

The objective of the management was to create a system where you either fit in according to their wishes or you leave. This is when I went back to reading *The Prince*, but not just as a story book, but attempting to find parallels in my workplace.

According to Machiavelli's recommendation, my first initiation was that I spent time not only examining the current state of my professional career, but also my previous years as a graduate student, which were far from satisfactory.

As I began reading and drawing parallels to my workplace, I used it to analyze my managers and professors and noticed how Machiavelli believed a prince should act. This analysis was not hard and was done by many an industry leader before.

However, most industry leaders wrote about Machiavelli from the point of view of how managers should control their teams. In my case, there was little I could do to influence the way my manager behaved, and futile about past managers and professors.

However, my analysis helped me to assign a pass or fail

grade to a manager. But here arose the problem—when I analyzed all my managers and professors, and assigned them a grade, almost all of them ended up failing, and a vast majority of them failed miserably.

What then was I to do?

I gradually realized this was a perennial problem plaguing most companies and there was little employees could do. They could attempt to solve the problem by moving to another jobs, hoping that someday they would come across that great manager, or by just giving up and branding their jobs as a waste of time.

Unfortunately, this affects skilled and talented engineers to a far greater extent than it affects the mediocre, as the former usually take jobs with great expectations and find themselves disappointed, while the latter are usually just thankful to have a job. It was clear that reading Machiavelli from the viewpoint as a manager or industry leader would not work for me as an engineer as I have no aspiration of becoming a manager anytime soon.

This is where I began to ask the question.

If an engineer weighs his or her manager or professor and finds him or her as mediocre or worse, incompetent, is that the end of the road?

It took many rereads of *The Prince* before I could think of a solution.

This is where Machiavelli as *Teacher of Evil* emerges to help an engineer trying to survive in a company. Machiavelli examined many strategies, some understandable and commendable, a few questionable and eyebrow-raising, and many that were downright evil and terrible, as he merely put together strategies that a ruler can use to rule effectively, without pondering about ethics or morals.

As already stated, in most companies, a regular engineer has little or no say in any decisions made, and therefore, if an engineer needs a way to survive, he or she cannot always use official channels.

Most of us were raised with an understanding of right and wrong and view the events of our lives in dark or light. However, the lower one is on the ladder, the light areas are extremely narrow, while the dark areas are much larger. If we are constrained only to the light areas, an engineer is in a straitjacket, and can only hobble around at the pleasure of the management.

Therefore, an engineer should look for grey areas to increase his or her freedom of movement. Put in simple words, if an engineer plays by the rules, all that can be done is follow their manager's instructions. Now, instead of playing completely by the rules, when an engineer attempts to bend or circumvent the rules without breaking them, the engineer's freedom increases.

Machiavelli writes in *The Prince* that if a prince lives only by how he ought to live, he will be ruined, and that though a prince would be expected to be purely virtuous, a few characteristics that appear to be vices if practiced correctly, can serve to rule effectively.

In the same manner, engineers expect their managers to be perfect, and in turn strive to achieve perfection, because that is how we were raised. A few vices in an engineer can lead to a successful career in even an extremely mundane job under the supervision of a very mediocre manager.

It took several years before I could use Machiavelli's writings to devise effective survival strategies that would work in many companies. These strategies need not be applied when engineers find themselves in very good jobs with challenging work and good pay. On the flipside, these strategies are also difficult to implement in one of those "jobs from hell".

Nevertheless, in most jobs, engineers find themselves under mediocre managers in mundane jobs.

When I finally decided to put pen to paper to write this book, I was convinced it has a purpose. While there are many books on leadership and entrepreneurship, I don't know of a comprehensive book detailing how to survive a '9-to-5 job'. As someone who has worked in 9-to-5 jobs over the past decade, besides also surviving a decade of being a graduate student, I am prompted by thought of having relevant experience, and besides, no one should wait to become a white-haired octogenarian before sharing their experiences.

So, for those who still ask this question "What is new in this book?", I answer, "It fills a void for engineers who want jobs where they apply what they learned and work peacefully without mental agony".

As already stated, Machiavelli's writings are complicated with many references to contemporary events of the time. Therefore, attempting to describe his book as he wrote it makes it difficult to understand.

After many rereads, I divided the chapters in *The Prince* into separate themes. Hence, a single chapter of this book will be based on multiple chapters of *The Prince*. Each chapter presents quotes, some of which are quite popular, and a few of that can be found online, but are not quoted very often.

Following the quote will be a description of his thoughts—not only about the quote itself, but on related paragraphs as it is neither practical nor proper to quote the entire section.

For those who wish to read *The Prince* independently, the reader will find it conveniently available in many forms, as the book is in the public domain in many countries. At the beginning of every chapter of this book is a list of the chapters contained within, so an interested reader can read those chapters first or even along with my book.

For those who still have their misgivings about how *The Prince* can be put to decent use, following every quote or extract, I draw a parallel to our modern corporate world from two angles. The first being about how managers use Machiavelli's philosophy and how these strategies can be detrimental to an engineer's well-being. The second is about how engineers can find loopholes in company management's usage of Machiavelli's philosophy to make decisions about one's employment under different circumstances and in different ways.

Inevitably, Machiavelli's philosophy is used to understand human nature to create the toxic workplace we are familiar with, although I present solutions that can be used by an engineer. My prime objective is to serve as a survival guide for engineers by examining their problems in detail.

Understanding leads to solutions.

Though Machiavelli's most famous quotes deal with wickedness, unfortunately, many of his lesser-known quotes are noteworthy for emphasizing the importance of wisdom, courage, and innovation.

In Machiavelli's analysis of historical events, it is clear that 'The Princes' whom he held in great regard were those who governed well and overcame tremendous obstacles.

He wrote *The Prince* as a gift to Lorenzo de' Medici—the ruler of Florence at that time—with the hope of getting an official appointment. The circumstances of his writing *The Prince* are by themselves worthy of an independent reading, though only a short description here is fitting.

Machiavelli held office in the Florentine republic that was created when the Medici family that had ruled Florence for sixty years had been expelled. After a decade of holding various posts, the Medici family regained control of Florence, and being accused of treason, Machiavelli was imprisoned and tortured.

When released, Machiavelli retired to his hometown where he wrote *The Prince* and a few other political treatises and notable plays.

However, his passion remained in politics.

In a letter to an old friend, he wrote:

> *When evening comes, I go back home, and go to my study. On the threshold, I take off my work clothes, covered in mud and filth, and I put on the clothes an ambassador would wear. Decently dressed, I enter the ancient courts of rulers who have long since died. There, I am warmly welcomed, and I feed on the only food I find nourishing and was born to savor. I am not ashamed to talk to them and ask them to explain their actions and they, out of kindness, answer me. Four hours go by without my feeling any anxiety. I forget every worry. I am no longer afraid of poverty or frightened of death. I live entirely through them.*

For reasons difficult to describe, the quote above lingers in my mind. I sense a wise and accomplished man who fell from grace, but still was able through his writings to forget his misfortune and put his wisdom to what use he could.

More importantly, through his writings, he found a measure of peace, and though *The Prince* was written with the hope of a return to public office, Machiavelli claimed to have no fear of poverty or death.

It is unlikely that these words were completely true, but finding peace and living without anxiety are things our world truly needs.

Therefore, for me, Machiavelli will always be a prophet of hope, and it is in this spirit that I present my book.

Table of Contents

The Modern Feudal System

THIS CHAPTER IS inspired by the following chapters of Machiavelli's *The Prince*:

- How Many Kinds of Principalities There Are, and by What Means They are Acquired
- Concerning Hereditary Principalities
- Concerning Mixed Principalities

In these chapters, Machiavelli describes the nature of different types of principalities as they existed in Europe in his time—and in the past. Initially, I mostly ignored these chapters and jumped to the next chapter which dealt with how principalities should be governed. However, later I realized their importance—they establish the basis and challenges of governance.

Machiavelli's book was set in the chaotic times of medieval Europe when democracy, human rights, and law and order—which most take for granted these days—were mostly non-existent.

At that time, Europe was a fiefdom with lords, dukes and other nobles ruling over tracts of land. The concept of nation states was taking hold and in Machiavelli's writings, one sees references to France, Germany, Italy, Switzerland and many others including smaller kingdoms such as Venice and Naples.

The common theme was major players conquering or ceding kingdoms to one another, either through war or diplomacy. Machiavelli's writings were shocking then and are shocking now. They not only describe tyranny but justify it as an essential component of governance.

However, in the historic feudal nature of Europe, it is easy to believe such tyranny was both practiced and accepted. In this chapter, I describe how the modern corporate world is similar to the feudal system of medieval Europe.

> *All states, all powers, that have held and hold rule over men have been and are either republics or principalities...Principalities are either hereditary, in which the family has been long established; or they are new. ... or they are, as it were, members annexed to the hereditary state of the prince who has acquired them...Such dominions thus acquired are either accustomed to live under a prince, or to live in freedom; and are acquired either by the arms of the prince himself, or of others, or else by fortune or by ability.*

During Machiavelli's times, there were republics and principalities. For a ruler, principalities might have a prince or a noble such as a lord or duke who in turn could be a part of a larger dominion under a king or emperor. Except for a few exceptional cases where a common subject became the prince through extraordinary ability and circumstances, ruling princes or nobles belonged to a dynasty.

The princes or nobles collected taxes from the common people in exchange for governance and protection. As stated by Machiavelli, the prince or noble might have his own armies or might use the taxes collected to pay mercenaries or auxiliaries.

A principality could be small or large and depending on the size, could have a single or multiple governing dynasties. The presence of multiple dynasties raises the question of who will be the supreme governor? Due to which, power struggles and violence were common. The governing dynasty could have acquired the principality through battle, be chosen through some form of diplomacy, or could be installed by a larger power. Principalities, with a new prince or noble, could also be new, possibly due to the division of a greater principality through a process of war or diplomacy.

It is easy to see the correlation with the world of business which is comprised of mega corporations with a multitude of subsidiaries—in contrast to mom-and-pop shops run by families.

The idea is that mom-and-pop shops are very different to the mega corporations, and this difference turns up in how they are managed.

The governance of a company could have a semblance of democracy or could be dominated by a founding family. Many small businesses—through a series of expansions—became major corporations we are familiar with. Furthermore, larger companies regularly acquire smaller companies or acquire divisions from other large companies.

In essence, similar to principalities that governed the lives of common people in medieval Europe, businesses govern the lives of citizens in modern times. Though one could argue that modern democracies with elected governments brought power to the common people, the essential method of providing sustenance for oneself through work is still in the hands of businesses and companies.

Machiavelli's writings were for the Prince and how he should govern, whilst this book is for the engineer wishing to survive and flourish in a company. Machiavelli wrote about common subjects who rebelled against their princes and how a prince could prevent such a situation from arising.

In the modern corporate world, an uprising of employees rarely occurs, though it is theoretically possible as often shown in television serials. In most cases, wishing only to be able to work in peace in a stable job, employees should devise survival strategies according to the nature of the company. Depending on company management and culture, survival strategies will differ.

The first step toward crafting survival strategies is to identify the problem which generally lies in company governance, which is good only in a few rare cases.

> *I say at once there are fewer difficulties in holding hereditary states, and those long accustomed to the family of their prince, than new ones; for it is sufficient only not to transgress the customs of his ancestors, and to deal prudently with circumstances as they arise, for a prince of average powers to maintain himself in his state, unless he be deprived of it by some extraordinary and excessive force; and if he should be so deprived of it, whenever anything sinister happens to the usurper, he will regain it. ...For the hereditary prince has less cause and less necessity to offend; hence it happens that he will be more loved; and unless extraordinary vices cause him to be hated, it is reasonable to expect that his subjects will be naturally well disposed towards him; and in the antiquity and duration of his rule the memories and motives that make for change are lost, for one change always leaves the toothing for another.*

Though some of the above is obvious, there are a few hidden inferences that are interesting.

In the case of a dynasty rule, where the heirs of the ruler naturally take over, the common people and many of the nobles gradually end up being loyal to the family name rather than the current ruler.

Even in the case of a dynasty rule, the beginnings might have been marked by turmoil and unrest. But if the founding rulers were astute and established a stable foundation, with time, the common people will gradually forget sufferings inflicted at the beginning.

This is what Machiavelli implies by: *and in the antiquity and duration of his rule the memories and motives that make for change are lost.* With several generations of dynastic rule, common people would enjoy a stable government for several generations, and one who has been born into peace and stability will see no need for change in the same way a wild animal born in a zoo will feel no urge for freedom.

Looking at some well-established businesses, particularly those dominated by a particular family, the initial turmoil during expansion will be replaced by the regular humdrum of business as usual. With many generations of salaried workers providing their services over their entire adult life, such jobs will continue to provide a secure income for many.

It is common to see several generations not only at the executive levels of such companies, but also among regular employees, as the older generations often take advantage of their positions to hire the next generations. Similar to caged wild animals, these employees see no existence outside these companies, though there may be other companies offering better positions and perks.

Before examining the governance of a "hereditary" company, we must first talk about what they are. In the business world, a significant proportion of businesses are family businesses. Though social structures differ vastly all around the world, almost every society has some form of loose or rigid class distinction. Within these class distinctions, there is usually a class that primarily undertook trade and commerce.

When we speak of modern businesses, most go back a few hundreds of years. The bedrock is that most commerce and trade were class-distinguished businesses that existed for centuries.

Though many companies that can be termed as hereditary have such roots to class-based business families, family businesses are no longer restricted to mere mom-and-pop shops that trade in basic goods. Even in these modern times, some of the mega corporations are controlled by just a few families. As required by law, many of them were forced to formally incorporate, which led to them being officially governed by a board of directors.

A closer look will reveal that most of the members of such boards are more often members of the founding family who by a majority vote, control the corporation as they please. Hence, family businesses have found a way to fulfil the requirements of the law while governing in a manner not so very different from before.

As Machiavelli infers in his book, one notices it is easy to maintain a hold over hereditary principalities. Thus, it is easy for an established family-run company or institute to continue. In the same way Machiavelli deduces, it would take extraordinary wickedness of a ruler of an established principality to earn the hatred of its citizens for him or her to be ousted. Similarly, it would take serious and long-term mismanagement of an established entity for it to be run to the ground.

Of late, we hear tales of iconic brands going bankrupt. A peek, however, tells tales of long-running mismanagement, refusal to adopt new technologies, poor finances and even scandals. The term "slow moving train-wreck" can describe these collapses.

Hereditary companies are usually process-driven and stick to what they have been doing. Since, in many cases, the founding families were well-established and able to expand their assets through careful investment, these companies have strong safety nets to withstand temporary downturns in the economy. The phrase *this is how it has been* is often heard. Those who have been around for a while might add a duration to it as well—*this is how it has been for the past twenty-five years as I remember it.*

Here, the communicator wishes you to know that your novel ideas can remain in your head—this old-timer has no intention of letting things change.

In some ways, this is understandable. An elephant cannot be expected to sprint like a cheetah. The prowess of an elephant is its formidable size and strength, and the elephant has given many a comfortable ride for many years. It is, however, important to understand that these entities—companies or institutes—are there for a reason, just like a forest or a lake. They provide sustenance to many.

Steady jobs afford comforting brands and helps manifest

secure financial and social status. At any bank, when you approach them for service and provide credentials of working at an established company, count the seconds before the bank official asks you if you need a personal loan. Consequently, in the same way as we are nowadays horrified at reports of deforestation or effluents released into a lake, attempts to shake up established entities are rarely appreciated.

These family-run companies typically are steady performers.

However, by mere law of probability, every family produces a brash heir or heiress who breaks the routine. Such a person is either a visionary who expands the business or, through sheer mismanagement, drives it to the ground. Many business analysts studying family-run businesses found an expansion cycle of a few generations—followed by a contraction cycle of another few generations.

However, due to the tight-knit nature of the governance of such companies, they are primarily affected by the persona of the members of the governing family rather than external factors.

With this being said about family businesses, let's talk about what it is like to be an employee in such a company. Here we depart from Machiavelli's philosophy, as we do not need a strategy to overthrow the governing family, but we need a feasible strategy for an employee to flourish.

As stated, in most of these family businesses, key positions in the company are held either by members of the family or those favored by the family. For that reason, management is usually comprised of family members or those close to the family. These people are usually well-settled, living comfortable lives and in most cases, without much burning ambition. For an employee to flourish in such a company, just create a favorable impression on the management.

There are various degrees of survival and flourishing within a company. By being loyal to management, one might expect to be well compensated. However, another form of reward, especially for an engineer, is to achieve a degree of technical success. Such forms of success can be in the form of fame within the engineer's domain or a degree of accomplishment.

In a family business, these can be difficult to attain, as rarely are those outside the family credited or allowed to gain preeminence. Though such a form of low-key employment which might be financially rewarding might have been sought after in the past, in recent times, most employees seek fame along with salary and benefits.

The means available to an engineer to seek renown—whether in a family business or otherwise—are similar and will be detailed later. In a vast number of cases, an engineer wishing to be accomplished must choose a path independent to that offered within the company.

That being the case, these means can be best described while also examining the state of employees in other forms of companies and businesses. To finish my thought about being employed in a family business, one could conclude that working in a family business might not be the most exciting of prospects, but the slow and steady nature of work offers engineers the peace of mind—making such jobs attractive.

> But the difficulties occur in a new principality. And firstly, if it be not entirely new, but is, as it were, a member of a state which, taken collectively, may be called composite, the changes arise chiefly from an inherent difficulty which there is in all new principalities; for men change their rulers willingly, hoping to better themselves, and this hope induces them to take up arms against him who rules: wherein they are deceived, because they afterwards find by experience they have gone from bad to worse. This follows also on another natural and common necessity, which always causes a new prince to burden those who have submitted to him with his soldiery and with infinite other hardships which he must put upon his new acquisition.

Here, Machiavelli talks not of completely new principalities, but those that have changed hands usually as the result of war or diplomacy.

In those times, rulers were rarely ever willing to live in peace. War was nearly continuous.

Therefore, many principalities changed hands, especially in Italy, which was fragmented. Machiavelli brilliantly analyzes the reasons by examining the example of the King of France invading Italy, but unable to retain his conquests. Before I extrapolate those specific reasons and liken them to the modern corporate world, there are many hidden insights into the analogy of a new principality in the modern economy.

In the corporate world, a completely new company is termed a startup. These are formed by one or more individuals with a feasible idea who decide to convert that into a product or service. This book will deal very minimally with startups for the simple reason that the author is not an entrepreneur and this book is not targeted at them. This book is for those who wish to navigate the corporate world as employees, and if at all, act as a reference to understand working context in startups.

In the corporate world, except for well-established brands where stability is the norm and most employees remain in their positions for decades, change is continuous and inevitable. Change appears in different forms. Machiavelli mainly spoke of changing a principality's ruler and its ruling class.

This occurs in companies during an acquisition or merger. Over the past decade, we heard of many iconic brands being acquired by another. In the world of software, acquisition is so common that it attracted the attention of legislators who fear monopolies of the big players. The acquisition of a startup by a larger and more well-established company is extremely commonplace, to the extent that it produced many workplace issues that are insufficiently addressed.

Though employees of a company are helpless when an acquisition occurs—either of or by the company they work for, acquisitions affect the life of even the junior-most employees. The most immediate effect of an acquisition is that the position of the employee can be made redundant. This can happen to employees working in either the acquired or the acquiring company.

Often, the purpose of an acquisition is to buy out a product, intellectual property, or customer network of the other company. Hence, after the acquisition, an employee in the acquired or acquiring company may lose their position.

Unfortunately, it is difficult to prepare for such a scenario as acquisitions and mergers are usually corporate secrets which only a few in the upper management are aware of. When an acquisition or merger is announced, there are usually only a few months before reorganization takes place.

Upon the first news of an acquisition or merger, it is advisable for an employee to assess whether a threat exists to their position. If a reasonable threat exists, a job search is advisable during restructuring interval. In many cases, a restructuring may not result in the termination of the employee, however having options will put the employee on better negotiating terms.

The risk to the position of the employee during the acquisition process has a deeper implication in these modern times.

As already expressed, with many iconic brands being acquired by larger companies, acquisitions are sought after in the tech world. Though celebrated in the business world, when an acquisition becomes the principal goal, it has negative impacts on the employees of companies that are on the acquiring spree—as well as those which are desperate to be acquired.

One could compare this to the modern education system, where students primarily study to perform well in examinations, and not with the objective of learning. The end-result, as we are aware, is that many graduates possess superficial knowledge, and though driven and ambitious, lack passion in their domains of specializations.

Most engineering companies have a technology or service that was the core of their growth. As companies grow, they expand to other domains and technologies. However, this expansion is not the result of internal engineering effort, but rather a process of acquisition, as many large companies usually prefer to acquire smaller companies rather than build technology themselves. This is due to the costs involved in a large company undertaking a project versus a smaller company doing the same.

Compare this to the effort required for an elephant to get up after lying on the ground versus a cat that decides to jump onto a couch after an afternoon siesta. As someone who has worked in mega corporations, a few conversations with senior managers exposed me to the disturbing logic behind this corporate behavior.

For a company driven primarily by profits, and most companies are, the decision about whether a particular service or technology should be in their range of offerings is a balance between the revenue it can generate versus the cost of providing it. When a large corporation enters a particular emerging domain of technology, this is usually the result of there being in the market a vastly greater demand versus the supply provided by existing companies.

This mismatch results in higher prices of technology and services, and therefore makes it financially rewarding for a large company to step into the arena. In such a case, due to the constraints of time and the fact that the large company may not invest in the technical capabilities needed to enter the domain, the only option is acquisition.

These companies then embark on a shopping spree like any of us going to the supermarket to buy groceries. Conversely, as demand subsides and revenues fall, companies sell off divisions that are no longer sufficiently lucrative. Ironically, those companies who avow themselves to be the champions of technology and place technology at the heart of all business, tend to be the most ruthless in acquiring or offloading divisions.

On the other end of the spectrum, smaller companies see acquisition as the greatest possible trophy and indulge in courting larger companies. With experience in a startup, I was often involved in management meetings about what would need to be done to be acquired by one or the other large company. This sometimes took on a feverish pitch—similar to teenagers constantly updating their profile on a dating website. Though it may not appear to be something one would expect to have a negative impact on the company, gradually the focus drifts from product development toward somehow just getting acquired.

In a company focusing only on acquiring others, the nature of work tends to become superficial and without significant depth as the eventual goal of the company is not to build technology, but to understand it to a degree that it can make a well-informed acquisition.

On the other hand, in a company desperate to be acquired, an engineer will find the scope and nature of projects continuously changing, as management jumps at anything that improves the prospect of being acquired. The uncertainty it brings an engineer can range from mere irritation due to lack of depth and focus on the work being done, to being constantly in fear of losing one's job.

With this continuous process of acquisition in large companies, an engineer can get frustrated—similar to a prisoner at a guillotine waiting for the blade to drop. In smaller companies, the threat to the employee may be far greater and needs to be assessed with respect to the nature of the management.

In some cases, management will provide employees with stock options, in which case an acquisition will result in the employees being compensated even if they lose their job. Sometimes, the employee will be let go with mere scraps.

Such scenarios are complex with each case being different and it would be best if the employee seeks legal help.

In larger or more well-established companies, changes occur for a number of reasons and acquisitions are usually not the reason. A change in management, similar to a change in a ruler, occurs when either the management of the company changes or when the engineer changes his or her job for various reasons.

In a vast majority of cases, the nature of the management plays the biggest role in the retention of engineers in a company. Sadly, management's intrinsic nature does not greatly differ between companies, and therefore, just as Machiavelli stated that *for one change always leaves the toothing for another,* a change in management rarely brings about a significant change in the working conditions of an engineer.

In theory, it's possible for engineers to oust a manager and replace him or her with another. Such scenarios are commonly depicted in television serials, they rarely occur in real life.

Most engineers have no control over who will be their manager, and any management change is usually a decision taken by the upper management. Engineers and other employees have to work under the current management or leave.

With the competitive nature of the modern workplace, changes in management occur often as managers change companies or transfer to other managerial positions within the company. To cite my example, I spent only two years in a large corporation and saw my immediate management change three times within that period. Therefore, a change in management is common in companies and an engineer needs adaptable strategies to deal with it.

The other reason for a change in management is because the engineer has changed his or her job. In these modern times, it is common to hear of employees either looking for a new job or being open to new opportunities after a few years of employment.

The reason for this change could be to escape a toxic workplace, to seek better opportunities in terms of better remuneration and benefits or better work assignments.

Even though changing one's job might appear a simple matter, unless the engineer undertakes this change with a proper mindset, the change either brings very little improvement or might make the engineer's conditions worse. This is also in line with Machiavelli's quote *for men change their rulers willingly, hoping to better themselves, and this hope induces them to take up arms against him who rules: wherein they are deceived, because they afterwards find by experience they have gone from bad to worse.*

Commonly, engineers who change jobs either do not see much difference in their working conditions or find themselves in a worse state.

This is usually because the engineer uses a single strategy to survive in different companies. However, due to the nature of companies and their different management styles, a single approach will not work in all cases.

This is where Machiavelli's philosophy can help. In understanding the nature of principalities and comparing these interpretations to the nature of companies. In completing this chapter drawing an analogy between principalities and companies, a few other forms of principalities as described by Machiavelli merit discussion. This is specifically because of the complexity of some forms of principalities and how close they are to the modern corporate world.

Machiavelli talks about how principalities with similarities in language and culture within a larger state that were ruled by a dynasty for several generations are likely to coexist peacefully with each other. This, according to him, pans out only when their laws are similar, and no drastic changes are made. Then, people will not strive for change to overcome minor differences in their benefits compared to their neighbors.

An impactful analogy is when schools and universities limit themselves to the few elites through very high fees and references. This is also observed in small family businesses where almost all employees are relatives.

Such institutions are well-funded, and with their students from very similar affluent backgrounds, would generally not feel the funding crunches or the need to strive for excellence. The teachers and the parents of students knowing their financial worth feel no need to pressure their children as they have their own resources such as family businesses.

As long as the children are kept out of trouble until they are absorbed into their own firms, they have little need to interfere in the institution's affairs. The institution knows that not much is expected of them and can therefore indulge in their own whims and fancies.

Machiavelli talks about the greatest challenges lying in the governance of a mixed principality with differences in languages, customs, and practices.

Almost every engineering company or university is of such a nature except for family businesses and elite institutions as cited above. In analyzing most cases, companies and educational institutions take on the role of mixed principalities where the leaders govern over faculty, students, and employees from very diverse backgrounds—both culturally and financially.

There are those with safety nets and who want either their degrees or paychecks and there are also those which are extremely ambitious irrespective of their backgrounds. In my educational preparation before a professional career, in almost every institution I studied at or thereafter in every company I worked at, management was indifferent to my holistic development.

On a positive note, it gave me insight leading to writing this book. On the flipside of the journey of my life, I had the opportunity to build a network and be amazed at how some people got to their positions, wishing they would've been my professors or managers.

Reading Machiavelli repeatedly gave me a new perspective into how and why institutions and organizations are managed the way they are.

The objective of this first chapter was to draw parallels with the feudal society of medieval Europe as described by Machiavelli to the modern corporate world in which we earn our living.

The problems faced by a ruler differ vastly with respect to the nature of the kingdom, and with their ambitions. A hereditary ruler choosing to live peacefully in their dominion had little to worry about—in contrast to the ruler who conquered a hostile state comprised of diverse ruling factions.

For that matter, a hereditary ruler could also choose to embark on a conquest like when Alexander the Great conquered the world, and therefore lose all the comforts that a hereditary ruler would normally have. The motives and actions of the ruler deeply impact almost every aspect of the lives of citizens, and therefore, even if we choose to declare ourselves as apolitical, that hardly provides an immunity from the actions of the ruling class.

A majority of the workforce may not have more ambition than to put food on their table, while a few might dream of being rich and famous. Several terms were concocted to describe the regular worker—salaryman, nine-to-fiver, working stiff.

Some look down on regular workers, while some take comfort in being a regular worker. Others resent being a regular worker and yet see no escape. Here too, one sees parallels to Machiavelli's medieval Europe where the regular citizens were little more than slaves, even though in theory they were free.

For a worker to survive in a company, the universally accepted strategy is to work hard and follow the rules. Unfortunately, this is rarely sufficient. If one plays a sport, one must understand not only the rules, but also victorious strategies. To survive in a company, one must understand the nature of the company and how it is governed.

As already expressed by Machiavelli, the greatest challenge lies in governing a newly acquired principality with different languages and customs.

Most companies fit the same mold. This makes it viable to apply Machiavelli's philosophy to the corporate world. Most of this book will be devoted to describing the nature of such companies and how an employee can survive in them. Most engineers surrender after working for a few years, as they believe nothing can be changed. I too almost succumbed. Except for a few rare and extreme cases, effective survival strategies can be formulated. An employee can flourish even in a badly managed company with meagre resources.

Why does all this matter to an engineer? When a company

collapses, everyone suffers—employees, shareholders and even third parties who did business with the company.

In my years, I saw companies go under and divisions disbanded—with all members either transferred or worse: fired.

The well-being of a company should be of interest to even the most junior engineer. Looking for a job when you are unemployed is never a rosy prospect. As a part of a company, most engineers are consumed by being the rat in the rat race mentality—just run with all the other rats.

One should remember that a rat is not a creature commended for its loyalty. Being a part of a rat race is not something to take solace in. On the contrary, the proverb *they left like rats leaving a sinking ship* refers to the fact that when times are hard, each must fend for themselves.

Anyone who has worked for a number of companies would be quick to attest that though there may be vast differences between managements, working in any company is rarely a bed of roses. So, even if one changes jobs, one can exchange one set of management-related problems for another.

What then is the solution?

Machiavelli explains. Defend a kingdom through arms and war as well as by statecraft and management. Many concepts can be translated into engineering companies. Most of us assumed Machiavelli's philosophy was designed for rulers and leaders. Per contra, it is surprising how well these can be adopted by every one of us and even more so by engineers.

As the saying goes—*it takes two to tango.*

A successful engineering company is not just the result of good management, but also of good engineering by competent engineers.

Conversely, a messy company with frustrated employees is not just the result of poor management but can be created by unimaginative employees unable to adapt and apply their skills.

Even if engineers have no power to change the management of a company, there are several ways to make their professional lives fulfilling. Even if their corporate careers languish, they can achieve a great deal independently.

The first step is to understand the nature of business and the challenges faced in making a profit, then devise a survival mode.

The Nature of Power

THIS CHAPTER PRIMARILY deals with Machiavelli's chapter called *Concerning Mixed Principalities* along with minor extracts from *Why the Kingdom of Darius, Conquered by Alexander, did not Rebel against the Successors of Alexander After his Death.*

Machiavelli's book begins in earnest with the chapter *Concerning Mixed Principalities, how they Should be Conquered and Once Conquered, how they Should be Retained.* Despite the chapter's subject matter revolving around military conquest, it barely describes military strategy. Instead, it covers political strategy in detail.

That chapter and the next few chapters are interconnected, complex and need to be read together and more than once as Machiavelli describes in detail many events that occurred in the previous hundred years or so when he wrote the book—with a few references to ancient Europe.

The chapter on mixed principalities describes the errors committed by the King of France when he attacked Italy and initially succeeded, but later lost all he had gained.

In comparison, Machiavelli describes how the Romans were able to not only conquer, but also retain their conquests. The chapter emphasizes the need for statecraft besides warcraft, laying the groundwork for some of his later chapters.

By the analysis of historical events, Machiavelli lays down the ground rules to be followed by a prince when conquering a mixed principality. Along with an extremely interesting historical narrative, Machiavelli throws in several beautiful quotes, many of which are not popular enough to be found on the internet.

The King of France, Louis XII, embarked on a conquest of Italy and found his first friends among the Venetians, who sought to use Louis to settle their rivalry with Lombardy. Once he got a foothold in Italy through the conquest of Lombardy, all the malcontents of Italy approached him to settle their quarrels with their rivals.

Given the fragmented state of Italy, this made it easy for the invading king to find support among most of the nobles and princes of Italy.

Thus, due to the ambitions of the Venetians, Louis controlled most of Italy with very little fuss. He could have comfortably stayed in control. However, he committed blunder after blunder.

To begin with, he assisted Pope Alexander to occupy a part of Italy called the Romagna. With this, the Church—which was a strong spiritual presence—now became, a formidable military presence in Italy. If an error of judgment of allowing the Church to enter was not enough, he further enlisted the help of the King of Spain to divide and conquer Naples. Eventually, the King of Spain drove Louis out of Italy.

While describing the actions of the King of France, Machiavelli does a detailed examination of what went wrong. To begin, the initial actions of the King of France were commendable as he entered Italy by taking advantage of the weakness of the Venetians.

However, from then on, he committed error after error. Instead of coming to Italy and establishing himself as the supreme ruler of Italy and disempowering the powerful factions and promising to protect the weaker factions, he allowed a powerful foreigner, the Pope, to enter. The rival factions now had someone else to ally with, and instead of him alone commanding all of Italy, he had to share the bounty with the Pope. The reason for this action was on behalf of a promise made by the King of France to the Pope, though, Machiavelli believes that a promise should never be the reason for a blunder, something which will be addressed in a later chapter. The final nail in the coffin was his association with the King of Spain, after which any hope of him remaining an arbiter in Italy vanished.

In comparison, Machiavelli praised the actions of the Romans, who avoided all the above mistakes. The Romans maintained colonies in all the principalities that they conquered and offered protection to the weaker factions while weakening the more powerful ones. Thus, the Roman empire flourished for long while extending over vast dominions. One might be quick to think of the Romans as nothing more than a war machine, but, from Machiavelli's analysis, in addition to warfare, they mastered the art of statecraft.

The references to the Romans that Machiavelli provided, however, are a bit difficult to follow independently, and therefore, one has to read the chapter for what Machiavelli implies rather than fact-check his writings.

The primary purpose of this chapter is to explore the nature of power—how those who seek it must behave in a certain manner to acquire and retain it, and how those who are unable to attain it need to find a way to survive.

In modern times, one talks about the two sides of the power equation—the oppressor and the oppressed, or the abuser and the abused, depending on the severity of the situation. Through the discussion of military and governing strategies, I extract what I believe to be strategies commonly used in modern times in the workplace.

The justification for these strategies will soon become obvious once we examine the historical facts that Machiavelli tried to analyze. I extend my assay with how these strategies can be dealt with in a form of antidote.

> *For, although one may be very strong in armed forces, yet in entering a province one has always need of the goodwill of the natives.*

The above statement sets the tone for the chapter by stating that a military solution alone is rarely sufficient for a long-term solution. A mixed principality—by its fundamental definition—consists of dominions that differ significantly in culture, and subsequently, is always ripe with rivalries and animosity between factions.

Whether such a principality is stable is a question of whether there is a single dominant power that binds them in the face of a foreign invader. Such a dominant power would need to stay friendly with other factions that are significantly powerful and instill a degree of fear in the smaller factions so they would submit, while also giving them a certain degree of assurance that they would not be oppressed to an unbearable extent.

When powerful factions are rivals and unable to form a consensus, the principality is vulnerable to fragmentation by a strong foreign power. While it might be impossible to attain unity among diverse factions, it is essential that the larger and more powerful do not take up arms against each other. This is obvious.

An invading power needs to exploit these differences and utilize the "divide and rule" strategy. In the case of a principality where powerful factions can be persuaded to switch alliances, a foreign invader finds it easy to gain a foothold. Though it may be just a few rebels that initially support the invader, many other malcontents will soon follow suit.

What follows next is then a question of how astute the foreign invader is versus how much force the ruling factions can raise. Machiavelli wrote many chapters and described historical events and there are many rulers he admired such that he described their actions in detail. What makes Machiavelli's writings informative? Despite praising many of the actions of some of these rulers, he still found some worthy of criticism, resulting in a balanced classification of his analyzed thoughts.

In our modern corporate world, the workplace is roughly divided into the management and the regular employees. In management, there is a division between those who own the company and the others who are strategic employees whose duties include controlling other reporting employees.

The objective of this management team is to maximize the profit of the company, thereby increasing the equity of the owners and in the case of those in the management team who are not owners, they expect bonuses or other similar forms of rewards.

The reporting employees in the company expect steady and long-term employment, with competitive compensation and benefits and reasonable working conditions. Except for a few cases, all successful companies need a stable workforce without too much attrition, for which the policies for the reporting employees need to be fruitfully content and engaging.

As already stated in the previous chapter, most companies resemble a mixed principality as employees will be diverse and from various backgrounds, socially, culturally and economically.

Exceptions to these are family-run small business where both the owners and a majority of the employees are related. In many other companies, even if management is dominated by a family, it is difficult to fill all the employee positions with family members and still be competitive. The previous chapter described how companies can differ in how they are managed depending on how much they are dominated by a single family.

However, further in the chapter, I will constrain myself to the case of a company in which both the management and the employees are diverse, thus resembling to the degree of closeness of the mixed principality Machiavelli describes.

Such a diverse company poses significant challenges both for the management as well as the employees. This is due to changing alliances within management, employees and between management and the employees without consistency present when either the management or employees are related by blood.

Additionally, with employees leaving or joining the company at all levels, besides promotions and cross-hiring, there arises the need to forge new alliances. Under these circumstances, there are many potentials for conflicts, with the management and the employees being subconsciously hostile to one another. Managers are fed a number of different strategies through various training programs. Many evolve their own strategies based on their experiences. Employees are also offered similar training programs, though these programs are aimed more at aligning them with the management goals.

Under these circumstances, a modern corporation resembles a medieval feudal society. A new manager needs a number of strategies to gain the confidence and loyalty from a team of engineers. These tactics, however, must continuously change as employees leave or join the team. Every exit or entrance has an impact on the team. To this extent, one will find Machiavelli's philosophy apt and easily translatable to a corporation.

On the other hand, hidden in Machiavelli's philosophy lies the key to survive, which is relevant to employees in a company. Survival strategies also differ widely depending on the capabilities of the employees. A poor-performing employee cannot use the strategies high performing employees have at their disposal. Also, it is inadvisable for a high performer to mimic what a low performer does to survive.

> Because, if one is on the spot, disorders are seen as they spring up, and one can quickly remedy them; but if one is not at hand, they are heard of only when they are great, and then one can no longer remedy them. Besides this, the country is not pillaged by your officials; the subjects are satisfied by prompt recourse to the prince; thus, wishing to be good, they have more cause to love him, and wishing to be otherwise, to fear him. He who would

*attack that state from the outside must have the utmost caution;
as long as the prince resides there it can only be wrested from him
with the greatest difficulty.*

Machiavelli believed that upon conquering a new principality differing greatly from one's own in language and customs, the best recourse would be for the king to reside in the new principality. This is obvious and can be said with regard to companies as well, though the globalized nature of business makes this difficult, if not impossible.

Management present among the employees will definitely be in a better position to hold their employees responsible for their goals and objectives as compared to a management that is either located remotely or frequently traveling. Previously, this was the case in any large corporation which might have offices in different locations in a country or even all over the world.

In recent times, with many businesses choosing to operate remotely either due to the period of the global pandemic or to cut costs, it is increasingly common for management to not be physically present among the employees and have a virtual connection.

Since the pandemic ebbed and economies returned to their pre-pandemic normal state, a major debate raged about whether businesses have the right to demand their employees completely return to the workplace, hereby ending all remote working privileges. For some businesses, such as factories and brick-and-mortar shops, remote operation was either impossible or caused reduced operations, therefore, a return to work was completely justifiable.

For a certain proportion of the workforce, a return to normalcy has resulted in a sigh of relief, as for them, their incomes had dwindled to unsustainable levels during the pandemic. For many others, it appeared as if the pandemic had either an insignificant impact on their revenues, or had in fact produced an expansion, and these businesses could decide to continue to operate remotely.

Yet, among these businesses, there were many who insisted that employees return to their offices. Various reasons were offered, ranging from increased productivity in a space designed for work, to an adherence to corporate culture which suffered during the pandemic.

A discussion on reopening policies is not applicable in this chapter. However, the purpose for visiting this topic is to examine the concept of management that is "on the spot".

If companies could operate effectively in a remote setting, this implies that the management continued to remain on the spot despite not being physically present. Here, one needs to revisit the phrase from Machiavelli's quote *"subjects are satisfied by prompt recourse to the prince"*.

From the perspective of an engineer, an effective manager is one who is closely engaged with the working of the team, and not necessarily one who is physically present in the office. More discussions will follow on various dynamics between a manager and their teams.

In modern times, active and engaged management can ensure a team stays engaged and motivated, as opposed to choosing a hands-off approach with disconnection from the employees.

> *The other and better course is to send colonies to one or two places, which may be as keys to that state, for it is necessary either to do this or else to keep there a great number of cavalry and infantry.*

According to Machiavelli, a colony would be the other option if the ruler cannot reside in the newly conquered principality. A colony would consist of those who have been dispatched to the new principality to govern on behalf of the ruler.

Of course, the effectiveness of such a colony will be decided on how capable and loyal the colonizers are. Withal, the objective behind such a colony is to assure the people of the conquered principality of governance that is equivalent to that in the conqueror's principality.

On the other hand, there still arises the need to perform certain unsavory actions such as to deal with potential rebellions upon conquering a new principality, which are best not done in the name of the ruler, but rather a colonizer. Machiavelli gives the example of Duke Valentino who upon conquering Romagna, found the need to appoint a cruel and ruthless governor in the form of Ramiro d'Orco to quell the unrest that had arisen from many years of weak governance.

Though d'Orco succeeded in restoring peace and order to Romagna, his cruelty caused him to be universally hated. To avoid this hatred culminating into rebellion, the Duke Valentino decided to have d'Orco publicly executed. This was carried out to persuade the people that the cruel actions of d'Orco were his alone and not the Duke's. In such a case, a colonizer can act as a buffer or a transition before the new principality is absorbed into the larger kingdom.

In the corporate setting, a colony or colonizers can be thought of as employees who take on management roles either officially or unofficially. When official, such employees have a certain proportion of management responsibilities in their contracts, as is the case of Team Leads or Senior Engineers, who though are not a part of the management, their key responsibility area is to supervise and empower junior team members.

In an unofficial capacity, it's common for well performing employees to end up being placed in charge of projects or other engineers, even though they were never contractually required to. It's observed that in some cases, the employees themselves choose to do so by walking the extra mile, while in other cases, they are assigned this responsibility by the management, either willingly or unwillingly.

A later chapter will describe some of these aspects in greater detail. At this point, anyone who has worked in an office will attest to the presence of these "seniors", who though not officially in charge, are considered to be the equivalent of bosses. These senior engineers act as a buffer when the management completely disconnects from the technical aspects of the company's operation.

The engineers have supervisors who have a technical background, and those managers do not have to deal with technical issues with their teams.

> But in maintaining armed men there in place of colonies one spends much more, having to consume on the garrison all the income from the state, so that the acquisition turns into a loss, and many more are exasperated, because the whole state is injured; through the shifting of the garrison up and down all become acquainted with hardship, and all become hostile, and they are enemies who, whilst beaten on their own ground, are yet able to do hurt.

It's thought provoking that Machiavelli made such an observation in those times of constant war and upheaval. One would have thought during those times that military occupation would be the order of the day.

Machiavelli in contrast, believed strongly in the military as will be delineated later, besides believing in governance and statecraft in separation from the war machinery. Except in certain rare circumstances according to him, where the ruler was an exceptional soldier and held the army in awe, a military solution to an administrative problem was rarely successful.

Though it was the task then of the army to conquer and secure borders, it was the task of the administration to provide a livelihood to the people and enforce the laws by which they could live and trade in peace. Coupled with that was the fact that during those times, human rights were largely unheard of, and armies could cause havoc to civilian infrastructure, the regular presence of the army in day-to-day life generally led to unnecessary bloodshed.

In the context of companies and universities, the equivalent of the army can be any entity whose responsibility is to enforce discipline, punish bad behavior, and in general to protect the company's objective of maximizing its profits and cutting costs.

In companies, such entities are usually their Human Resource Department, while in universities, there could be several committees and bodies for this purpose. In academia, most institutions typically have only punitive measures to control their students, as though offering financial rewards may not be feasible, most would not even choose to bestow other privileges, as it would only complicate the overall governance of the institutes.

Unfortunately, one gets conditioned to the state where one rarely receives any reward for good performance, but is promptly punished for poor performance. In companies, such an approach will rarely help, though many companies would prefer to choose such an approach, for the simple reason that it would cost the company less than a rewards-based approach. This results in increased attrition, but to ascertain whether the levels are alarming would be dependent on a number of other factors.

Referencing the last three extracts earlier cited from Machiavelli, it might appear to be disconnected and similarly the same can be said about the comparison to the corporate world, where the primary question here is—who will rule over the newly conquered mixed principality?

It can be direct, indirect or through brute force.

There is an assumption on the part of Machiavelli that either direct or indirect rule will be effective. Given the fact that ineffective rule could result in a rebellion that might end in the execution of either the ruler or the colonizer, there was ample reason for either the ruler or the colonizer to make the best possible effort at governance.

In the modern corporate world, as one is painfully aware, poor or mediocre managers are most often not only ever present at almost every level, but last for an agonizingly long time in various different positions. Therefore, there is almost an immediate contradiction in the comparisons between Machiavelli's philosophy and the modern corporate world—a ruler must possess a certain degree of ability to be able to conquer another principality, whereas managers in many cases are mediocre.

How then does one apply the philosophy in this case?

It is important to remember that the conquering ruler has his own kingdom in good order, keeping the people content and maintaining a well-trained army. In conquering another principality, and especially a mixed one, it is not necessary to achieve the same degree of rigor in governance.

The reasons are numerous and will be described in the later chapters as well. In the simplest case, it may not be advisable to drastically change the entire system of governance immediately after the conquest, as such changes will injure people to a degree that they may find ways to resurrect their old king and rebel against the new ruler. Rather than a drastic and immediate change, a conquered principality can be gradually imbibed into the larger kingdom over a span of several years, by which time, with peace and prosperity, the conquered people now identify with their conquerors instead.

Until then, all that needs to be achieved is to ensure that the conquered people do not rebel, and with trade and commerce, the people are content. In such a case, the ability of either the conqueror or the colonizer is not tested to the same degree as in his own home country.

Let us now translate the above concept into the modern corporate world. As already stated, most managers themselves are employees and are also moving between organizations or within the organization. Therefore, the managers do not have to behave in a manner as they would have been expected to, were they also the owners of the company.

The difference in the dedication of the manager when an owner or a stakeholder in the company versus a mere employee is stark, though, it doesn't immediately imply that those type of managers would necessarily be effective managers.

However, in the case of a manager who is merely an employee of the company, the primary objective of the manager is to hold on to his or her position, just like any other employee of the company, whereas for a manager who owns the company, the drive to operate the company profitably is much higher.

In the case of such a manager who is simply another employee of the company, the comparison to a ruler who conquers another mixed principality is apt. For this manager to retain his or her position, the goals and objectives of the team under his or her supervision need to be fulfilled to a certain degree. This is similar to a conquering ruler, whose prime objective is to remove all possibilities of a rebellion, but yet at the same time, allow the people of the conquered principality to continue to live their lives with a certain degree of normalcy.

In the case of the conquering ruler, it would be wise to allow the conquered principality to gradually be absorbed, while, for the manager who owns no stake in the company, it is not his or her responsibility about how the company is managed, especially in the long run.

With this comparison between Machiavelli's recommendations of how a ruler who conquers a mixed principality should behave and a manager of a company who has no ownership of the company in terms of stock options, we can elaborate on the three broad strategies that a manager can adopt.

The first is for the manager to be directly connected with their team, follow closely all operations and maintain an open channel of communication for good as well as bad feedback news.

Though it might be the obvious and only solution, it is difficult to find a manager who can achieve this.

This is partly due to the fact that when employees embarks on a managerial career path, they might wish to disconnect from the technical aspect of operations. The reasons could be many, with compensation and benefits playing the primary role, since managers tend to be better paid than engineers and also receive larger bonuses based on their team's performance. Other reasons include a greater opportunity to travel, interact with clients, and potentially relocate. Often, the reason for choosing management over core engineering is an inability or unwillingness to keep up with the constant challenges of evolving technology. Therefore, though an engineer would ideally like to be able to discuss everything with their manager, including technical issues, often this is simply not feasible, as the manager could have chosen to not actively be involved in engineering.

It is however extremely important that a responsible individual with technical skills be accessible to the engineers and is also able to hold them accountable. It is unreasonable to expect mere engineers to devote themselves completely with no supervision to bring their projects to a conclusion.

In the absence of supervision, one can expect either engineers to while away their time and let projects flounder or become frustrated with the lack of support needed to be able to complete their projects and with time, eventually leave the company.

Neither of these outcomes are desirable for a company.

For this reason, if the manager wishes to disconnect from technology, it is extremely important that they appoint an effective engineering supervisor. Depending on the size of the company, this can be achieved either through company policy, or by the management carefully grooming engineers into such supervisory positions.

Several large corporations have distinct management and technical growth paths, where those who wish to remain closely connected to technology can continue to do so, while still growing through the ranks of the company. The rungs on the technical ladder are various levels of technical supervisors—from a team lead to a technology leader.

By providing such a distinctive path, such companies can ensure that engineering supervision is available at every level, and any technical person at any level always has access to someone in times of need and is also accountable to someone who has knowledge of technical matters.

How rigorously a company can groom and promote talented engineers through this ladder, of course, vastly differs.

In some cases, unfortunately, these technical positions are either filled by those without sufficient technical skills or by those who are unable to supervise others.

Of course, that is after all not the only aspect to a manager surviving in a company. A manager can do all the above and still fall prey to either other managers, or to a superior in the upper levels of management. This book will not deal much with management, as the book is targeted towards engineers, rather than managers. However, it is safe to say, that when an engineering team flounders, and is unable to fulfil its goals and objectives, it is very difficult for a manager to survive. It would take great social and political skills for a manager to survive despite their team performing poorly.

Machiavelli has also spoken about a few cases where rulers who deviated from the norm, were still able to retain their kingdoms. Some of it will be covered in later chapters, as some of the philosophy is translatable to engineers as well.

Again, the prince who holds a country differing in the above respects ought to make himself the head and defender of his less powerful neighbor's, and to weaken the more powerful amongst them, taking care that no foreigner as powerful as himself shall, by any accident, get a footing there…

To illustrate the above, Machiavelli differentiates between the King of France who did not observe this rule and on the flipside how the Romans who had observed this rule.

Any foreign power that wishes to invade a kingdom needs the initial support of a few of its principalities. Typically, these are those who have grievances against their ruler and look towards the foreign power as their liberator. The foreign power must defend and protect these malcontents, yet at the same time not allowing them to gain greater power.

As for those who were powerful and content with their ruler, they need to be weakened so as not to eventually seek to overthrow the invader. With this power balance between those who were weak and those who have been weakened, the invader must then ensure that no foreign power as strong as the invader must ever gain a foothold in the conquered land.

This ensures that those who have been conquered do not flock to the new invading power.

For many of us who are not students of history or follow politics intensely, the above seems logical. However, when one stops to think of how closely this philosophy is applied in modern day corporations, one would probably wish they had spent more time learning history.

To understand the implication of the above in the setting of the corporation, this extract from Machiavelli which is more abstract will send a shiver of an awakening through anyone:

> *He has only to take care that they do not get hold of too much power and too much authority, and then with his own forces, and with their goodwill, he can easily keep down the more powerful of them, so as to remain entirely master in the country.*

For any engineer who had to hold the short end of the stick due to favoritism or nepotism, the above statement will make it clear why that might have happened.

In any corporation, the ones with power bestowed upon them are those in management. One could argue that everyone in a corporation possesses some form of power, but to start with a simple pondering argument, it is the managers who are in power, and by one acknowledging the definition of this power, it is understood to be the capacity to direct or influence the behavior of others or the course of events in a corporation.

As already stated, this complex case of Machiavelli where a ruler needs to conquer and retain a kingdom that is culturally and linguistic diverse, is the norm in any corporation. Therefore, for a manager to gain control of their team or department, the methods to be followed are very often similar to Machiavelli's belief in how a foreign power should invade and retain a kingdom.

This discussion will be more pertinent in the case of managers who are mere employees rather than those managers who are also part owners of the company. This is due to the fact that a manager who owns a part of the company will be far more diligent in the operations of the company, and will attempt to extract a greater deal of commitment from team members.

On the other hand, managers who are mere employees do not stand to gain significantly if the company generates a profit, these managers are usually content as long as they retain their positions and receive bonuses. This kind of manager has goals, objectives and targets that need to be met, which they have on promotion been assessed upon.

Unless the company is in an extremely competitive domain, meeting targets only requires a manager to be reasonably competent and not have a team of shirkers. Unfortunately, most managers use the bandwidth available to them to form cliques in their team and pitting one faction against another.

Now the question arises, which group of employees will the manager favor and which group will need to be curtailed? Social media overflows with posters of the ideal manager who encourages the diligent and disciplines the slackers. Anyone who has worked in a company will usually laugh when they come across such a post, as the reality is so very different from the idea.

Just as Machiavelli can be understood here that, in order for the manager to command their team, the mediocre need to be reassured that their positions will not be threatened by those who are talented, while the talented need to be curtailed to ensure that they do not outshine their manager.

Sadly, this is the exact opposite of what one sees on social media—or even what one might expect if one thinks logically.

Most of us have seen and experienced associations like this in innumerable ways, wherein—some are overt and extremely destructive, others sly and minor irritants. *How often have we seen the best projects and best resources assigned to the mediocre while the talented are left with dregs? How often have we found mediocre employees receiving high praise for completing ridiculous and trivial tasks while the talented will receive not a word of acknowledgment for even a stellar performance? How often have we found grave errors on the part of the mediocre overlooked while minor errors committed by the talented receive the spotlight?*

As long as the mediocre are not hopelessly incompetent, managers are willing to provide them with leeway.

I encountered this form of behavior from managers and know of countless many engineers who experienced it as well. Incredibly frustrating to any dedicated engineer, most of us blame this on vain and petty managers who pander to mediocre flatterers.

Though shallow human nature does play a role, this strategy is deliberately adopted by managers to ensure that they have full command over their teams. By raising the mediocre to a pedestal, they establish that level as the baseline of the team's performance. By curtailing the talented, they ensure that the talented will never dominate the team.

Just what Machiavelli advised.

In all my years working in different companies, I used and saw many other engineers use various strategies to cope and minimize, if not completely avoid the frustration caused by the management. Some take a combative approach, gallantly fighting tooth and nail with the management. Some take on a dismissive role, deciding that their jobs were just not worth their time and energy, and minimizing the effort to decrease the inevitable disappointment. A few unfortunates get drawn into the current, and end up being miserable and at certain instances even resorted to self-harm. The coping mechanisms used are strongly correlated with the personality of the engineer, which is similar to how a human being approaches life in general.

As someone who tried out different coping strategies before finally settling on things that worked, the next few paragraphs will describe some of these strategies and the context in which they failed or succeeded.

Let us first examine the case of the fighter—the one who fights tooth-and-nail to prove themselves. Social media loves these kinds of engineers: the ones who not only possess technical skills, but also are passionate enough to fight. In my experience, very few engineers fall into this category to begin with, as the time and energy to gain expertise in technical skills usually leaves little time to invest in understanding the nuances of human behavior and being able to fight against those who have spent a majority of their time in gaining social and political skills.

How often has one come across a skilled engineer with equally commendable communication skills and social skills, successfully navigating the cliques that exist in the workplace?

The immediate answer is: rarely.

This led to many engineers spending a significant portion of their time and energy on gaining "soft skills". Though soft skills now encompass almost everything that is not directly related to core engineering and includes other useful skills such as project management, accounting and finance, the original definition of soft skills was to augment the technical skills of a person with social skills such as communication, social interaction and human ethics.

Everyone can agree that the ability to speak clearly and communicate effectively with a diverse group of people from different backgrounds is a very useful quality for anyone to have, and includes engineers and non-technical people.

However, the soft skills engineers were pushed to acquire are generally targeted towards either molding them into managers or into employees that will be the favorites of the management.

These skills are provided to engineers under the guise of helping them integrate into the culture of a corporation. However, in most cases, they are designed as brainwashing tools to change the mindset of an engineer into normalizing management practices that are not in the best empowering interests of the engineer.

Sadly, I spent a lot of my precious time in these training programs before realizing they were wild goose chases. As an engineer, all I wanted was to work in peace on interesting projects and not get involved in management practices. This was also the case with many other engineers whom I met in my career, and I found they felt the same way when forced to take up corporate training programs.

With or without training in soft skills, very few skilled engineers gain the favor of the management to be on the better end of nepotism—or avoid it altogether. Most would find their frustration only increasing, as with increasing investment in undergoing training, they find themselves unable to deal with the management.

A few become combative and aggressive only to find themselves further isolated and labelled as troublemakers. An apt proverb goes thus *"do not argue with fools, a passer-by may not know the difference."* For a skilled engineer, outmaneuvering management is an almost impossible task, which, if even possible, would require an effort so great that it would rarely be worth it.

Let's examine the second coping strategy, which is a general dissociation from their jobs.

Over the course of my career, I observed many engineers frustrated to the point that they described their jobs as a complete waste of time. Many divert their energies to other pursuits—family, friends, travel, social causes. Though these activities can decrease the emptiness one feels when a large part of their day is spent idling away their time, eventually the feeling that their technical skills were never valued can cause a great deal of agony. For a skilled engineer, the greatest trophy is to be accomplished in their domain and accepting that they achieved nothing during their careers is a very bitter pill to swallow. Therefore, though this coping strategy can bring temporary relief and happiness, in the long term, it usually results in disappointment.

To surrender and give up on one's career and happiness should not be an option.

Sadly, I saw many engineers resign themselves accordingly.

Such disconnection is inevitably followed by an unhappy professional life. It may appear that the situation is completely hopeless with the engineer condemned to a life of frustration. There are many who accepted this and lost all hope of professional satisfaction.

When I began my industry career, I came across a few who preached this like a gospel. Though my initial experiences in the industry were frustrating, I could not resign myself similarly the way some had chosen to. It took many years and a multitude of experiences before I was able to craft a coping strategy that worked.

There were also a few engineers who for some reason continued to believe that their work would be recognized, even though they found themselves repeatedly sidelined. Some were too naïve to even understand that they were being sidelined, while a few others clung on to the hope that their hard work would eventually bear fruit.

The beliefs of those mindset engineers were subconsciously at many times in my analysis closely linked to them being religiously devout, syncing the commonality of scripture that *"the meek will inherit the earth."*

Whether these engineers attained what they hoped for was something I never found out, as I did not remain in touch with many of them. However, observing the way these engineers were being treated was agonizing to the point that I would never put forth such a strategy as a solution.

Before I describe a potential coping strategy, let's step back and look at the big picture, as in modern times, we are continuously brainwashed into accepting a multitude of blatant lies as New Commandments.

Machiavelli spoke objectively about conquest and statecraft, and the objective of this book is to translate his philosophy into success and achievement in the corporate setting. Therefore, one must delve into the quintessential nature of modern work, to examine it in the most basic form, what it means to work and what we expect from our work.

From this, let's delve into the essence of power hierarchy at work, what it means to be in a position of power and what it means to be subservient without power.

Having a job in the simplest form is to devote a proportion of one's time and energy in completing a set of tasks that are assigned. Over the past decades, almost every job transformed from its erstwhile relatively simple form to one which in modern times has become so convoluted, that any employee is subjected to what might be equivalent to a death by a thousand cuts.

Without diving into the details of the unfairness of most modern-day employment contracts, every aspect of the above simple definition of performing a job has been twisted against the employee. In the simplest form, it has been either an extension of the time expected by the employee to devote to the job, or at the other extreme, a complete erosion of working times resulting in an expectation from an employee to be always available.

The other aspect of a job consists of the tasks assigned to an employee, which as most of us have experienced has also become clouded in uncertainty nowadays. Most of us have experienced constantly changing requirements or projects launched, then scrapped on the fly.

Add to these, in some situations, employees are expected to devise tasks and take ownership of projects. In my years in the industry, I have come across all of these—projects changing every few weeks, or to be assigned projects with no instructions or background, but expected to deliver results.

For most engineers, this scrambling to figure out what to do can be frustrating as one does not expect to concoct tasks when they would expect that to be the job of the management.

One can find a multitude of books and training programs that advise employees on how they should deal with different scenarios. During the early days of my career, I used some of the strategies suggested by my company's training program and found them to be of limited use.

Beside their ineffectiveness, I found that the need for an employee to accept this randomness as the new normal and stretch and contort oneself to be aligned to the modern work environment, would only result in most employees burning out. As long as the management felt the employees would continue to adapt to whatever obstacles they placed before them, those obstacles would only continue to multiply.

The root cause behind an employee's continuous struggle to clear every obstacle placed before them, is the inherent human need for those at the weaker end of the power relationship to attempt to gain the approval of those at the stronger end of the power relationship.

Conversely, the assumption of those at the stronger end of the power relationship is, that those at the weaker end must bend over backwards at their whim and fancy, this then acts as the driver of most of the irrational expectations from management at the workplace.

Unfortunately, most of us are unable to view the workplace as an entirely different arena from our family and other personal relationships. Additionally, most of us see our jobs as extensions of our educational institutes and universities, where professors and teachers are considered even in these modern times as mother or father figures.

When one works in a company, it is important to realize one leases their knowledge, time and energy to achieve the goals of the company. One must also realize that a company's reason for existence is to make a profit for the stakeholders, which could be the owners and other investors.

Unfortunately, most of us come to our jobs with false hopes, either expecting the company to readily fund every project and idea that might come to our fancy or expecting a company to pay no heed to profit while chasing excellence. I have during my career heard the naïve phrase *"they care only about profits, and have no appetite for challenges."*

Conversely, with profits being the primary driving force for any company, an employee who focuses his or her energies on the generation of profits for the company will find little difficulty in retaining their position in the company.

As employees, we find ourselves at the receiving end of many expectations, many them ridiculous. Every manager expects employees to devote 100% to their jobs. There are some who will do so blatantly, while others will do so smoothly. A job could begin with an employee committing only the required regular hours but could gradually transform into one where the hours become long, and even weekends are not free.

In most cases, this increasing pressure at work is usually not due to a genuine increase in work, but rather a reassignment of work. This reassignment of work occurs due to the factors mentioned above, and those who find themselves burdened with meaningless work are usually the skilled and talented. The exception to this trend is usually if the company is in an extremely competitive domain, in which case, the mediocre are just a few and the talented are many, and if employees are burdened, it is usually with interesting work.

In such a case, if the employees are well compensated, they usually have very few complaints.

It is usually a rare case to find a company where the talented form the majority, the work is genuine, and compensation is commensurate. In most companies where this is not the case, there can be for short spells a similar situation, where the management is willing to put aside its usual policy of favoritism towards the mediocre and give the talented a free run.

This is usually the case during stressful periods before

appraisals or reviews, where a team might be forced to deliver results. The management will shamelessly exhort the talented few to use any resources available, to work long hours, if necessary, to meet targets. I have experienced this numerous times in my career and am quick to identify this behavior as a byproduct of weak management.

In a vast majority of cases, the work in a company is mundane and rarely challenging for a skilled engineer.

The reason is obvious.

Engineering programs in universities are continuously raising the bar on the quality needed to be achieved to graduate. Companies in turn, can pick well-trained engineers, even though the work they have to offer requires nowhere near the level of training achieved.

With profit being the primary driving force in any company, only a few can afford to invest in new product development or for that matter research. This mismatch between skills available and quality of work conducted is a stumbling block that will never disappear, as universities do not train their students to fulfil the requirements of industry, but rather maintain their own independent standards.

It is obvious that this disconnect between an engineer's expectations and a company's requirements will lead to a number of chronic problems.

From the company's perspective, they need to mold these engineers into future assets for the company. The definition of an asset differs vastly between companies. For a relatively small company that may not have a vast range of products or services, an expert is just someone who is well-versed with the company's products and services, in the technical aspects and also with the other commercial aspects.

In some companies, engineers are given a degree of freedom in project execution and are well-compensated. In such cases, if the engineer is not over-worked and if the job offers opportunities to travel and interact with clients and other third parties, even if the technical aspects of the job are mundane, the overall experience is positive.

In many companies though, the experience is negative for reasons that could range from inadequate pay, excessive pressure, unbecoming clients and in some cases even poor working conditions.

For a larger company with a wider range of products and services, an expert is a potential technological leader who has experience and knowledge of several domains. In my short stint at a few global multinational companies, I came across a few such people who had decades of experience in several domains in the company.

These people were always very impressive, highly accomplished and took a great deal of pleasure in their work.

During the orientation of new employees, it was customary to invite some of these experts to motivate fresh employees. Though it was obvious that several decades of work in a large company could result in an accomplished and satisfactory career, it was clear that not everyone survived the marathon.

Whether in a small company or a large one, the survival rate of employees depends largely on the management of the company. Here we can reference the earlier discussion about the management technique adopted that is typically pitted against talented engineers.

When we consider the chasm between the skills and training of most engineers and the technical requirements of companies, the possibility of a successful engineering career is out of reach for most engineers for the simple reason that a prosaic professional life allows little possibility for accomplishments.

Thus it is no surprise that the survival rate of employees, particularly the talented ones, is very low. These factors turn the career of an engineer into a minefield.

However, in exactly the same way Machiavelli laid down the basic principles to be followed by a ruler, it is possible to translate that philosophy into a basic set of guidelines that an engineer can follow. In attempting to translate Machiavelli's philosophy into a survival strategy, it is necessary to tear down some of our most accepted beliefs and practices, and instead attempt to discover what needs to be done, rather than what we were told to do.

The past few pages were dedicated to describing in detail some of the challenges and obstacles faced by an engineer. The next few pages attempt to find gaps in those obstacles in exactly the same way one would sweep a minefield.

The very first obstacle is as already described as the large gap between the technical requirements of a company and the skills possessed by an engineer. This gap exists in almost every company, even the massive global companies where normally one would not expect that to be the case.

The reason is that every company tries to recruit the best possible engineers, and therefore, even in the case of a Fortune 500 company recruiting graduates from ivy league universities, the gap still exists. Very few companies work on cutting-edge emerging technologies and can afford to narrow the gap until it is almost non-existent.

At first glance, this is a death sentence for most engineers.

To overcome this obstacle, we need to reflect on the basic definition of a job, which is to lease one's skills, energy and time to achieve a set of tasks. When one expands on this definition of a job, one places obstacles before oneself.

Unfortunately, an engineer is brainwashed from various sources into a narrative that will continuously expand any job definition until it is completely unsustainable. Social media is full of unrealistic stories about the role an engineer should play in a company, from being the active owner of every project to being the driving force in a company.

Anyone who has worked in a company will quickly dismiss these stories as unfounded. Therefore, our first step forward will be to determine what is the absolute minimal requirement that needs to be met by an engineer to survive in a company.

An engineer in a company will be assigned a variety of tasks.

Though an engineer may label themselves a purely technical person, a fair proportion of the tasks assigned to the engineer may be non-technical, either related to project management, finance, or customer relations. This is due to the company objective to make a profit. Profits are generated by the successful transfer of products and services. Hence, the company expects the engineer to perform tasks necessary for it to generate a profit.

There are many cases where an engineer may be dissociated to a certain extent from the profit-making core of a company, which is usually the case with working in R&D or advanced product design.

However, even in these cases, the engineer will still have goals and objectives that eventually align with the profitability of the company. While an engineer might be quick to get frustrated with this requirement, there are several aspects of profit-making that can actually be used to the advantage of an engineer.

In a vast majority of cases, companies make a profit by either mass-producing goods thereby decreasing cost while still commanding reasonable market prices, or by offering services that are repeatable and streamlined to decrease the time and cost to the company.

To ensure profitability, companies focus on decreasing variability and increasing repeatability. Though companies need to regularly expand their range of products or services in order to remain competitive, this expansion is still performed ensuring that the cost incurred is eventually recovered through profits.

It is often that this focus on repeatability results in many engineering tasks being repetitive and turning out mundane.

The disadvantage to engineers is that the lack of challenges and interesting assignments makes them feel their training is being wasted.

On the other hand, the advantage is that for a skilled engineer, the effort required to perform the assigned tasks to a fair degree of thoroughness is low. Therefore, if one places survival in a company as the sole objective, with most companies being profit-driven, one expects the survival rate to be high.

It is clear there is a glaring contradiction.

Though it should be an easy matter for any engineer to survive in a company, for various reasons, most struggle to do so. Therefore, it is an easy conclusion that there are strategically placed obstacles to ensure that engineers struggle to thrive in an organization.

To understand these obstacles, we already examined how managers use the weapon of favoritism to put down talented engineers, in exactly the same way, an invading power needs to curtail powerful factions in the conquered land. A manager has at their disposal a vast array of tools to obstruct and hinder the progress of engineers.

Nonetheless, though it may appear a lost cause to overcome these obstacles, in reality it is not too difficult, as most of these obstacles can be ignored and bypassed.

Let us begin with the very first tool used by managers, which is favoritism.

As previously stated, managers use favoritism to create factions within their team to promote a group of people in the team who will later support them, thereby resulting in a symbiotic relationship.

To guarantee that the chosen group will support them, that group must be dependent on the manager in turn. As a result, these kinds of managers will usually choose a group of mediocre engineers to form their support base, as these are the ones who are least likely to survive without the active protection of the manager. This group will usually be favored in various ways for instance, they would receive the best projects, it would most often appear that they would be least burdened with unnecessary activities, most often they would be allowed to speak and present first in meeting and events, and more than often would be showered with praise at almost every occasion.

As already stated, the usual strategies of being combative, dismissing the job altogether or trying to turn into one of the flatterers, is unfeasible for a talented engineer.

Much the same, there are a multitude of solutions suggested by numerous books and training materials devoted to producing the ideal corporate employee. In most cases, these solutions are inappropriate for talented engineers, as the social skills and maneuvering needed to get on the greener side of pasture is usually not something skilled engineers possess or are willing to devote their time to acquire.

Before labelling this obstacle as unsurmountable, one should ask the question, *is this even an obstacle that needs to be overcome?* In my experience, these obstacles can be safely bypassed without jeopardizing one's position in the company.

Aforementioned, a job is an exchange of skills, time and energy for compensation and benefits. For a talented engineer to indulge in office politics and join factions is a waste of time. By attempting to join a faction, an engineer only ends up being a pawn in a game played by those who lack technical skills. When one finds oneself in an office rife with factions and with the management encouraging and pitting groups against each other, it is advisable to restrict one's efforts to the basic tasks needed to perform one's job to a reasonable degree of competency.

To put in one's best effort under these conditions will rarely yield any appreciation or praise, as such workplaces are notorious for work being stolen and credit being withheld.

If one feels that by avoiding office politics, one hinders career advancement, the following quote from Machiavelli is apt:

> *From this a general rule is drawn which never or rarely fails: that he who is the cause of another becoming powerful is ruined; because that predominance has been brought about either by astuteness or else by force, and both are distrusted by him who has been raised to power.*

The mediocre are mere pawns used by leaders to maintain power and those who rise to power will continue to play their power games. Unless one wants to join this rat race, it is best to avoid it altogether. Most leaders assume the talented will continue to give their best for whatever scraps are thrown to them. It is for the talented to have faith in their talents and abilities and choose their own paths wisely.

Aside from favoritism, another tool used by the management is to burden engineers with a variety of non-technical tasks. Most companies have several processes that all employees need to follow, and these could range from basic project management and project accounting to requiring them to be involved in other activities.

As previously stated, the prime objective of any company is to generate a profit, which requires essential tasks to be repeatable. When employees repeatedly perform tasks, they gain expertise through repetition whilst errors continuously decrease. As the time required to complete tasks with a lower probability of errors decreases, the cost incurred by the company decreases.

For any company, these essential tasks are the most important, as they have a direct impact of the profitability of the company. The other tasks that have been assigned to employees are usually not critical, and in most cases is due to the fact that the company would prefer to decrease their costs by not hiring more employees.

From all the myriad of tasks assigned to an engineer, it is important to identify those that are essential and those that are not. For the survival of the engineer in a company, all that matters is that the essential tasks are performed to an acceptable degree of thoroughness. The remaining tasks can be either delayed or performed in a casual manner without too great a risk to one's position in the company.

In later chapters, I describe another adaptation of Machiavelli's philosophy which will describe how an engineer should project oneself in an organization. Notwithstanding, in terms of survival in a company, it is obvious that a talented engineer should not have too much trouble as the requirements in most companies are basic.

From the above discussion, the best advice for an engineer struggling to survive in a company is to fulfil the basic requirements that are critical to the company. For engineers struggling in toxic workplaces where managers are constantly sidelining them or denying them credit, the best solution is to not seek praise or commendation in those companies.

Be that as it may, indulging in office politics in the hope of being a recipient of any favors bestowed on the group that one is a part of will rarely ever yield results, as one will only turn into a pawn. It is deemed by health experts that indulging in workplace politics is both frustrating and futile, as politics is usually the bread and butter of those who possess little technical capabilities and therefore devote all their energies to it.

An engineer thrown into workplace politics is equivalent to jumping into a rattrap, and should be avoided.

The above discussion focused on the survival strategies for an engineer in a company. However, for a talented engineer who spent a great deal of time and energy in acquiring those skills, mere survival through performing the bare minimum of essential tasks will rarely be adequate.

For every talented engineer, the greatest joy is always in executing challenging projects. We chase every carrot dangled in front of us in the hope that it will be that fascinating project which will define our careers.

Alas!

We engineers should accept a hard truth—most of these carrots will amount to nothing. To put our talents to use, an engineer must separate themselves from the organization. The organization exists to make a profit and their position in the organization requires them to work towards that organizational goal. For leverage, an engineer must seek challenging projects independently.

One of the best parts of this modern world we live in is how interconnected it is. Challenges exist everywhere and collaborators can be found all over the world and projects that started with little other than an idea have gone on to be household names. For every engineer, my suggested advice is: *open your mind to challenges outside your organization.*

A later chapter will describe this in greater detail.

To this, I find this quote of Machiavelli absolutely apt about the Romans:

> ...*nor did that ever please the Romans which is forever in the mouths of the wise ones of our time: Let us enjoy the benefits of the time—but rather the benefits of their own valour and prudence, for time drives everything before it, and is able to bring with it good as well as evil, and evil as well as good.*

No rewarding career is built in taking comfort in a shallow position merely because ones' immediate needs are met. We need to acknowledge that things will always change. The benefits one was accustomed to will gradually be cut or even worse—be taken away. If you are talented and capable, have faith in your ability and talents and use them to the fullest. Choose goals carefully and to whom you will assign your loyalty.

In this light, another quote of Machiavelli stands true:

> *The wish to acquire is in truth very natural and common, and men always do so when they can, and for this they will be praised not blamed; but when they cannot do so, yet wish to do so by any means, then there is folly and blame.*

Hereinafter, it is suggested as fruitful to push oneself to the greatest and achieve the most that one can as an engineer, which will bring satisfaction and happiness.

Acting out of desperation and jumping into every wild goose chase hoping that it will yield great rewards will result in a state of unhappiness and frustration.

To conclude the above discussion, an engineer should separate survival from accomplishment. To survive sustainably in a company provides a degree of financial and social stability, and except for a few of those who come from wealthy families, to achieve financial independence is a great leap forward.

Caving in and allowing oneself to be frustrated by a toxic work environment and throw away opportunity is unfortunate, and sadly I have seen many do so. To find satisfaction in utilizing one's talents, engineers needn't constrain themselves to the company, but rather seek to participate in side-projects. Besides finding professional independence, a side-project can turn into a lucrative opportunity, besides also allowing one to network with like-minded professionals.

Before concluding this chapter on how an engineer can deal with managers who will use a variety of tools to retain control over their teams, it is interesting to describe another observation made by Machiavelli on the different types of power structures.

In the chapter where Machiavelli describes how different types of countries are ruled, he makes a comparison between Persia and France.

Persia was ruled by Darius at the time he was defeated by Alexander. Darius was the supreme leader of Persia and every official appointed in every post within the country was his servant who he could dismiss or transfer at his pleasure.

On the other hand, France was a collection of principalities—each with its own lord, The citizens of that principality owed allegiance to their lord. The lords in turn owed their allegiance to the King of France.

However, the French King was not widely acknowledged as the supreme leader of France as each principality had its own power structure.

In conquering Persia, Alexander overthrew Darius in the battlefield. Conquering one or more provinces on the border had little effect as everyone swore their loyalty to Darius and these provinces were eventually taken back by Darius.

Nonetheless, once Darius was defeated and killed by Alexander, Persia was won by Alexander and they then recognized Alexander as their ruler.

Therefore, even after Alexander's death, Persia was held by Alexander's successors if they maintained their relations between each other and avoided war.

If France was conquered, things would have been completely different. One or more provinces could be captured by an invading ruler and if the lords submitted to the invader, those provinces would become theirs. As history documents, the King of France could put together alliances to eventually drive out the invader or limit him to a few of the border provinces as the rest of France had their own power structure.

To win over a significant portion of France, an invader would have to use the methods stated before—befriend a few of the weak, increase their power to overcome their rivals yet limiting their power so that they stay loyal to the invader.

Let us compare this to the corporate world.

You could say there are small companies versus the mega companies. Small companies have limited employee sizes managed by the owner. All acknowledge the owner as the one and only true source of authority with any intermediate doing their bidding.

In comparison, a mega company has a complex power structure where the leader who might be the director, CEO, President or Chairman is a distant figure, who, though is acknowledged to be the leader of the company, does not feature in the regular affairs of the team. The team acknowledges the immediate manager and the few levels who are superior to them.

In a small company, an engineer needs to primarily keep the owner happy. In comparison with the impression an owner has of an employee, the impression of others has little impact even if they are senior. A good impression secures the position of the engineer, and it would take a prolonged period of poor performance to reverse their standing in the company.

A bad impression is difficult to correct and such an engineer has a limited future in the company even if the others have a good impression. Unfortunately, many capable engineers do not understand the dynamics of human relations, and even if they do, are unable to control them to their advantage.

Human relations are difficult to control, but to push them in a favorable direction is easy. To begin with, the first impression is, if not the last impression, at least a lasting impression. Setting a good first impression—even if it means not being yourself or impersonating a character very different from your true self—goes a long way to consolidating your position in the company.

However, in a larger company, things get more complex and impressions are hard to create, but then again impressions can also change. A manager is not the owner of the company and has in turn to report to another manager or many levels of management.

Therefore, under varying pressure, their behavior to engineers will change. Moreover, the impression created by an engineer at a point of time usually only lasts for a while and in some competitive organizations with engineers vying for the favors of a manager, a wily manager will strategically change their behavior to their team of engineers.

In such circumstances, there is usually a time cycle to most changing patterns. To hold a good impression for a sustained period, it is necessary to sense critical times and behave in a way that creates a good impression. On the other hand, it would take a duration of poor performance for an engineer to reach the point of demotion or dismissal.

Which of these is better is tough to determine. Having worked in both types of organizations and having created good and bad impressions which have changed over time, every workplace has its complexities. It is important for an engineer to remember: *you are not responsible for the overall wellbeing of your organization.* Corporate culture is usually such that it will tend to dump greater responsibilities on employees and even mere engineers will sometimes be pressured into being responsible for projects.

As an engineer, you are truly responsible only for the

technical aspects of the projects. Any other responsibilities assigned are usually to maintain pressure. Therefore, assign priorities to tasks and focus on utilizing technical skills. Citing earlier as expressed, an engineer should not rely on the company alone for technical challenges. If the tasks in the company are menial, then let those menial tasks be done well and seek challenges elsewhere. If the essential tasks assigned are done in a manner that shows diligence, one's position in the company is protected as much as it possibly can be.

Though the two chapters of Machiavelli from which chapter is drawn are short, they contain a wealth of information about strategies that can be employed to gain control over human beings in general.

Machiavellian strategies an invader should employ are actively used by many managers in controlling their teams of engineers.

These strategies can cause a great deal of frustration and agony to engineers, and most of the well-publicized solutions do not help, but in fact only sugar-coat bad management strategies or worse, brainwash engineers into accepting these management strategies as the noble truths.

Engineers who survive these nefarious managements evolve their own survival strategies after being burned numerous times.

Using a bit of imagination, it is possible to interpret Machiavelli's suggestions to an invading prince into survival strategies for an engineer in an organization.

The aspirations of an engineer in an organization can be divided into major arenas—the first being survival so as to be able to provide for oneself and one's family, and the second being to be able to utilize one's education and talents to achieve success and gain renown.

This chapter lists reasons why it may not be possible achieve these goals in an organization.

The reasons range from the profit-driven nature of companies, as also the huge chasm between the educational qualifications of graduating engineers versus the technical requirements of most jobs, and the petty nature of most managers, who lack technical depth and are usually content with surviving in their jobs by any means.

This chapter describes how, if one throws away all the social media narrative and focuses on the essentials, the profit-driven nature of companies can make survival in a company easy for an engineer.

As for all the strategies used by management against engineers—nepotism, cronyism, over-burdening employees—one can either choose to not involve oneself in these games, or one can intelligently be diligent in only those tasks that are necessary for one's survival in the company.

Of course, this chapter raises the question—*what about those who do not wish to stop at mere survival?*

The answer to this question can be found in the later chapter dedicated to this, titled *Starting from Scratch*.

The Urge to Destroy

IN MACHIAVELLI'S CHAPTER called *Concerning the Way to Govern Cities or Principalities Which Lived under their Own Laws Before They Were Annexed,* Machiavelli sheds light on another aspect of modern life I could not relate to among other of Machiavelli's chapters.

In this chapter, Machiavelli expresses options available to a ruler who conquers a principality that had either lived in freedom, or if not completely autonomous, had their own laws and local government—similar to a republic with elected officials and a decision-making body comprised of its citizens.

Machiavelli believed a principality which had autonomy and the freedom to create its own laws should be handled differently from a regular principality under the control of either a prince or a noble or was part of a larger dominion.

The chapter is subtle, and on first read, I found no analogy to the modern world. To introduce the relevance found in this chapter, let's start with a quote from *The Prince*:

> *Whenever those states which have been acquired as stated have been accustomed to live under their own laws and in freedom,*

there are three courses for those who wish to hold them: the first is to ruin them, the next is to reside there in person, the third is to permit them to live under their own laws, drawing a tribute, and establishing within it an oligarchy which will keep it friendly to you.

Machiavelli says that when a ruler takes over a principality which was accustomed to being governed in a particular manner by an established ruler, there are few options—the first is to destroy them, the second is to reside there and govern or to establish an oligarchy which will keep the people friendly to you.

The advice to reside in a newly acquired principality or use trusted leaders is understandable and was discussed in my previous chapter—if the new ruler is ever-present, any disturbance or rebellion can be quickly noticed and put down. If not, the advice to establish an oligarch is also a good one as this oligarch is one of the people, well-respected and if that person accepts the new ruler, the chances are that the people will eventually accept the new ruler with time.

If the ruler is one who conquers often, the first option to reside or send trusted leaders may not be practical as the ruler cannot be in every conquered territory or send trusted to those colonies. Finding an oligarch who will accept the new ruler may be a long shot, since if anyone has money and power, that person may eventually want to take over the new principality for oneself and end up leading the rebellion against the new king.

Now comes the advice to destroy them.

Machiavelli states that when a principality lived under a ruler or a dynasty for a long time or were living in freedom as in the case of a republic, they get accustomed to it. When a new ruler conquers them, they long for the old times. Those who are powerful might plot the overthrow of the new ruler and if not put down promptly, could start a rebellion.

This is where Machiavelli states that it might be best to simply destroy the newly acquired principality—disband the powerful among them, remove many of their established structures and offices and erase their identity. Machiavelli gave examples where some who have conquered have dismantled their conquests and succeeded in holding on to them, while others who did not dismantle their newly acquired territories ended up losing them.

> *And he who becomes master of a city accustomed to freedom and does not destroy it, may expect to be destroyed by it, for in rebellion it has always the watchword of liberty and its ancient privileges as a rallying point, which neither time nor benefits will ever cause it to forget. And whatever you may do or provide against, they never forget that name or their privileges unless they are disunited or dispersed, but at every chance they immediately rally to them…*

In the times of medieval Europe, wars were violent and rulers could subject their citizens to terrible cruelty, who in turn had little recourse to any justice.

There were many instances of invaders who completely destroyed cities which they conquered, reducing entire settlements to rubble, executing all those whom were deemed a threat, without in many cases even a sentencing, let alone a trial.

Such actions in these modern times would constitute war crimes and international courts have tried politicians, soldiers and others accused of war crimes. Though this book is not about modern politics, one can say that with such actions now being difficult if not impossible, it is not surprising that most conquests in recent times have gone badly, and the conquered have in many cases reacquired their freedom.

As already said, though we might live in a democracy, our workplaces for the most continue to be autocratic and those company managements do still have a significant bandwidth in maltreating employees.

During my years as a graduate student and as an engineer in companies, one aspect of the behavior of professors and managers always puzzled me.

In many cases, I sensed they acted out of pure malice and went about with a metaphorical sledgehammer. Cases of indiscipline apart, in many cases, these destructive acts were carried out against skilled students or employees. Being at times a witness as well as a recipient of such destructive intentions, I wondered what was passing through that boss's mind as the behavior transcended shooting oneself in the foot, though on the hand it rather was equivalent to committing self-destruction.

Hence, when reading what Machiavelli wrote about how principalities who lived for a long time under a ruler should be governed, I have some inkling about what those professors and managers might have been thinking.

So, here comes the comparison with the behavior of professors and managers who behaved like marauding raiders. Lamentably this arises from basic human nature: *people are scared of the unknown.*

Some professors think a bright student might put them down at some point decides to preemptively cut that student down, or a manager decides a talented engineer might outshine them, and so takes them off all challenging projects and instead dumps on them mundane tasks. This thought process stems from the need for the leader to destroy a subject who they think has no loyalty towards them yet can rise up against them.

In many cases, actions such as these are completely uncalled for, but these actions—unfortunately—feed the dark side of unethical human nature. Moreover, the manager believes that by destroying a talented employee, at a later stage, with even a few scraps thrown at them that, they in return will remain loyal and humbled, as measurably the employee is receiving something for their sustenance.

Destructive behavior is common in academia where professors have the protection of tenured positions and they do not fear repercussions from their bad behavior as long as it is not completely egregious.

Moreover, the license society offers most teachers to behave as they please produces a toxic environment. During my years in academia as a graduate student, I found most professors behaving in a destructive manner at some point of time. Some felt the need to use destructive behavior to assert their authority, whilst others carried on a tradition of random behavior. A few others enjoyed causing destruction.

Destructive behavior is also very common in large companies. Several studies were conducted in recent times on toxic workplaces and the effect on employee mental health. The fact that so many studies have been reported is an indication of how widespread is the problem.

One expects harassment would be less likely in large companies with well-established human resource departments, having put in place processes to deal with complaints. Unfortunately, in many cases, not much will be done to address the root cause of these complaints, as the primary role of the human resource department in most organizations is to protect the company rather than the employee.

Therefore, complaining employees often find themselves going around in circles as the complaint resolution process is a means to divert attention from the bad behaviour.

To understand the root cause of destructive behavior such as harassment, let's examine how and when Machiavelli suggested a ruler destroy a newly acquired principality.

The need to destroy arises from the need to gain control over an entity that may either be strongly loyal to another power, or may have been free and therefore, might be unwilling to submit to power.

In a large corporation or a university, students and employees come from diverse backgrounds. The corporation or the university would like to change the mindset of these diverse groups of people to make their primary affiliation towards the company or the university, rather than any other social structure.

As the students and employees gradually lose their individual identity and consider themselves solely as students of the university or employees of the company, they become pawns to the power structure.

This erosion of identity can take place through either reward or fear. One could argue that reward and incentive might be a better option to gain the loyalty of students and employees. However, fear is a cheaper weapon. To create a culture of fear, only a few need to be empowered with the task of harassing others. In some cases, managers and professors play the role of the harassers, while in others cases, the harassers are those given a free hand in what could be said to be a case similar to state sponsored dissonance.

The purpose of this form of harassment is to make the student or employee primarily fear the company or the university and erode any other loyalty they may have. The effect is that students and employees sacrifice their families, personal lives, and even their own individual identity to retain the positive impression of their managers and professors.

Employees take pride in working long hours almost to the point of burn-out, just to wear the tag of the stellar employee. At this stage, the company turned the employee into a slave who can imagine no other existence outside the company.

One might wonder, *how would any such environment even be functional?*

Surprisingly, many such companies and universities can not only survive, but also thrive. In the case of large corporations and universities, these entities already have the advantage of size and branding, which need significant mismanagement to run them to the ground. Therefore, this culture of intimidation results in a few students or employees exiting, which is a small price to pay.

In the long run, by producing brainwashed employees, a company achieves far greater gains. As for the risk of complaints and litigation, few of them will even see the light of day, as most students and employees will suffer in silence than risk their degrees and careers. In solution, all that universities and companies invest in are rigorous disciplinary structures to eliminate discontent among students and employees at early stages.

As someone who has been in a few toxic environments and survived, the solution I suggest offers a workable strategy in many situations. There are of course, extremes, such as the case of the manager or professor from hell, with whom it is impossible to work. In a vast majority of cases, one needs only to pretend to live in perpetual fear of the management for their harassers to feel they have achieved their goal.

In a later chapter, this trait will be discussed in greater detail while discussing Machiavelli's philosophy of how rulers should keep their word. However, all a company wishes to do is mold employees into worker ants who imagine the company is their universe. Such a pretense is easy to put forth as it is much easier to feign timidity than to feign bravery.

Pretending to be the model employee a company wishes to break one into can be achieved against almost every tactic used by the company. As an example, many companies will impose almost inhuman working hours onto new recruits. Except in very competitive domains where the long hours translate into equivalent work, in most other cases, the long hours are presentism—appearing to work hard in front of the managers and colleagues, and sacrificing one's personal life for the company.

In any company, it is impossible for an employee to be constantly supervised and their work to be scrutinized. Companies are using innovative techniques to spy on their employees, and some have even resorted to installing spyware on the computers used by some employees to monitor their activities and detect periods of inactivity which could imply wasted time.

Nevertheless, most of the strategies used by companies to pressurize employees focus on preventing employees from idling away their time. As long as the employee appears to be genuinely engaged in some form of work, these strategies would rarely be able to detect the subtler forms of diversions.

The best tactic to be used in such cases is that of salami-slicing. This is when the employee gradually diverts away parts of the official working time into activities independent of the company. There are many such activities an employee can indulge in that would not be considered as a violation yet can be used to enhance their professional profile.

As long as the employee does not violate confidentiality or indulge in activities that are a direct conflict to the business of the company, an employee can choose a number of side projects that can either result in gaining a new skill or create a new potential source of revenue. Therefore, even though an employee might be spending long hours in the office, a significant proportion of that time could be used for the employee's own gainful activities.

Another form of harassment meted out to employees is a deliberate strategy of assigning random and varying tasks with the intention of browbeating them and breaking them into the *"corporate culture"*. Management is fully aware of the human need to invest a certain minimum amount of time and energy on a task in order to be able to complete it. Therefore, the constant barrage of assignments is a technique for harassing an employee into living under a constant fear of disciplinary action.

As already stated, unless a company is in an extremely competitive domain where there is indeed a great deal of genuine work, assignments are manufactured to burden employees. With experience, employees can distinguish between genuine work and concocted work and assign priority to tasks.

All the same, the objective of the harassment strategy is to coerce the employee into living in fear, and all the employee must do is pretend to live in fear. With time and experience, a reasonably talented employee will be able to perform tasks to a reasonable degree of competency without too much effort.

Also, most companies deliberately create an atmosphere of fierce competition to pressurize employees, and by constantly comparing their performance, create an atmosphere where employees are jostling each other for the favor of the management.

It is important to note that when you are being pushed aside, you are also being pushed out of the spotlight. As a star employee, your every action is observed and noted and you are always under pressure to perform. As a dark horse, you have much greater freedom to work in peace without being scrutinized constantly.

Therefore, being the dark horse is a great time for an engineer to begin a side project or accelerate an existing one. To give up and get frustrated once one realizes that one has lost the favor of their manager, is a very common reaction for many engineers, and this is precisely what that manager wants. If one works quietly on their side projects while giving their manager the impression that they have succeeded in being broken, one will be actually augmenting ones' skills and knowledge which the company otherwise could not have offered.

As a consequence, when the time comes, when that manager pretends to *"give another chance"*, one would have nothing to lose by graciously accepting that offer while letting that manager believe that you are returning them such, with your tail between your legs.

It may be necessary to adapt survival strategies according to the nature of a company on the whole to deal with destructive intentions. Remember that these strategies are *intended* to erode the employee's spirit.

At the same time, one might argue that such strategies should be dealt with through laws and legislation. Understand that most organization are adept in finding ways around every law.

Labor laws continuously evolve through the centuries and though the conditions of workers have drastically improved in comparison to the beginning of the industrial revolution, exploitation of workers continues and will continue to take place.

Human greed is a quality that will never disappear, and one needs to find ways to adapt and survive. Hence, in addition to skill, effort and perseverance, one should adopt a degree of deceit to survive.

The purpose behind this chapter was to describe how sadly, the darker side of human nature tends to align with corporate strategy to sometimes make the conditions of employees if not unbearable, extremely unpleasant.

At first, many of us attribute it to megalomaniacal managers who think they are entitled to behave how they feel. Per contra, in examining Machiavelli's recommendations of destroying the spirit of people who enjoyed freedom, this behavior has a purpose rather than merely the result of bad character.

Indeterminately to survive under these circumstances, one only needs to understand the true purpose behind why such companies feel the need to harass its employees.

Later chapters will cover some other aspects of workplace harassment when Machiavelli describes historical events and how they led to the downfall of some rulers.

Starting From Scratch

THIS CHAPTER IS based on the chapter written by Machiavelli titled *Concerning New Principalities which are Acquired by One's Own Arms and Ability*.

In this chapter, Machiavelli speaks of entirely new principalities founded by those who rose to the position after being private citizens.

Machiavelli describes a few princes who founded new kingdoms and it is clear from his narration, he had the greatest respect for any prince who did so.

> Let no one be surprised if, in speaking of entirely new principalities as I shall do, I adduce the highest examples both of prince and of state; because men, walking almost always in paths beaten by others, and following by imitation their deeds, are yet unable to keep entirely to the ways of others or attain to the power of those they imitate.

As Machiavelli stated, a new prince, and especially one who is not accustomed to being in power, has only the examples of past princes that he can use as a guide. Nevertheless, danger lies in the fact that those kinds of princes may not be able to mimic their actions completely and may not achieve what their idols achieved.

Thus, for a new prince to be successful, he must not only seek as guides those before him that he considers to be great, but also learn to improvise when and wherever necessary.

Machiavelli gives the examples of a few such leaders in history such as Moses, Cyrus, Theseus, Romulus and Hiero as examples of those men who from a private station, achieved the title of prince.

Among these, of course, Moses is probably a mythical figure, yet Machiavelli draws a parallel between Moses and the others, as almost all the others mentioned followed a similar path. To begin with, was the opportunity—that these founders were among people who were oppressed and looking for a savior to lead them to freedom.

Citing Moses, he found the people of Israel oppressed by the Egyptians; Romulus who was abandoned when born in Alba, and not remaining there, contrarily founded Rome; Cyrus found himself among the Persians who were discontented with the rule of Meses; Theseus on the other hand found the Athenians dispersed; and Hiero whilst found the Syracusans oppressed.

> And in examining their actions and lives one cannot see that they owed anything to fortune beyond opportunity, which brought them the material to mold into the form which seemed best to them. Without that opportunity their powers of mind would have been extinguished, and without those powers the opportunity would have come in vain.

Here comes the realization of the delicate balance between fortune and ability.

One should be grateful when finding success.

However, fortune in these cases provided the opportunity for them to display their ability, and it was the founding prince's own inherent ability and astuteness that eventually decided the outcome of history.

> *And it ought to be remembered that there is nothing more*
> *difficult to take in hand, more perilous to conduct, or more*
> *uncertain in its success, then to take the lead in the introduction*
> *of a new order of things.*

Though the above statement might be obvious, its significance is so great that it is worth emphasizing. The allure of doing what everyone else is doing and simply following the rat-race is extremely strong. In consequence, one must remember, that evolution itself is based on the concept of *beat the competition* followed by *survival of the fittest*.

Therefore, the origins of entire species of living beings are founded on the principle of a few that diverge from the normal path, avoiding the competition for resources, and choosing to enter new domains. To point out, this choice need not always result in a new species, as to survive under new circumstances, thus will result in many perishing in a process called natural selection. This process of natural selection yields the strongest to form the new species, which as we all know is how humans eventually evolved from the microscopic living beings that first inhabited our Earth.

> *The difficulties they have in acquiring it rise in part from the*
> *new rules and methods which they are forced to introduce to*
> *establish their government and its security…Because the*
> *innovator has for enemies all those who have done well under the*
> *old conditions, and lukewarm defenders in those who may do well*
> *under the new.*

In establishing a new kingdom, the new prince must establish new laws, create a new army or dismantle an old disloyal army and establish a new one. Along with that, he must establish public offices and a system of government. These may not always be welcomed, as those who are powerful might prefer to live in anarchy where they can profit the most, and few would pay taxes willingly, if they could rather live in complete freedom.

Machiavelli says this is the domain of great men and women, and those who have succeeded have their place etched in history. It requires great skill, bravery, intelligence and confidence to establish a new principality and govern it. Those who oppose will be strong and will have no doubts, while those who support might be unsure and of dubious capabilities.

Therefore, to turn the tide and create a system with a strong foundation, governance will need a leader to rally around him those with ability and maintain their faith in him.

> Hence it is that all armed prophets have conquered, and the unarmed ones have been destroyed. Besides the reasons mentioned, the nature of the people is variable, and whilst it is easy to persuade them, it is difficult to fix them in that persuasion. And thus it is necessary to take such measures that, when they believe no longer, it may be possible to make them believe by force.

Here, once again, Machiavelli throws a shocker—as another example of how Machiavelli's writings are supportive of tyranny, he considered this to be a necessity in most circumstances.

Throughout history, one hears of visionaries being punished and vilified. Some may secretly admire these rebels and what they gave humanity, yet would still never choose that path. For this reason, Machiavelli says that if the rebel is a man of peace, he will rarely succeed, as humans are variable, and will change their loyalty as their need arises.

Thus, to depend on goodwill alone is impractical and the rebel must use force, cruelty and deceit where necessary, to ensure that the new order is successfully established, and those who oppose it are put down. This is a concept I rather painfully accepted as the truth, and in several places in the chapter, I refer to it.

A comparison to establishing a new principality in the

modern world is obviously the case of entrepreneurship. Many household names we now swear by were a few decades ago mere side hustles that began in someone's basement or garage. It needed great skill, perseverance and hard work to convert those projects into sustainable companies.

Those who started these companies are of course well known and they are all over the news and social media—*Elon Musk, Steve Jobs, Mark Zuckerberg and many more.* Entire books were written about them and many quotes became almost engraved in stone.

Though, for many reasons, one might have cause to revile them, one cannot deny that they have made their mark on humanity and left behind companies that will last for decades. Along with these, there are also the stories of many scientists and innovators that are noteworthy, and this book will mention a few.

I strongly suggest to engineers, that whatever may be their domain of interest or specialization, spend some time on reading about those who are noteworthy in ones' particular domain. Focus ones' read on how they started, how they persevered and if they faced hurdles or temporary setbacks, to analyze them as well.

As Machiavelli puts it all too beautifully:

> *A wise man ought always to follow the paths beaten by great men, and to imitate those who have been supreme, so that if his ability does not equal theirs, at least it will savour of it.*

As already stated, this book is not about entrepreneurship as I am not an entrepreneur. Nonetheless, there is an aspect of entrepreneurship that I have actively imbibed and would recommend all engineers to do so as well.

In the past few chapters, I briefly described how side-projects and independent activities can be undertaken by engineers to separate themselves from their companies and achieve a fair degree of success independently.

In this chapter, I will examine the various aspects of this in greater detail, and how even engineers with no aspirations of becoming entrepreneurs, can occasionally wear the entrepreneurial hat to transform their careers and their personal lives. It is my sincere endeavor through these chapters to empower continued reading, as original thought and action bring about many surprising benefits that are not immediately evident.

A lot of the discussion in this chapter is specific to engineers, as I will recount my experiences and describing the experiences of a few others whom I have either studied or been acquainted with. Notwithstanding, anyone in any domain can be just as innovative as the process of innovation is an attempt to improve existing processes.

Hence, I will leave it up to the reader to adapt the contents of this chapter according to ones' domain, and even in the case of probability that ones' domain differs from engineering, a relevance of empowering thoughts is available for self-analysis.

To begin with, it is important to describe what is meant by innovation, as it is a term that is intimidating. Most working people distance themselves from it. For most working people, the foremost goal is to achieve financial stability and the risk associated with pursuing esoteric objectives is simply not feasible.

Thus, a quick conclusion is that innovation is the pursuit of either those with advanced degrees or those who have financial and social safety nets to risk a failed enterprise. Though many ground-breaking innovations required a great deal of time and energy besides many other sacrifices, the process of innovation can be practiced continuously with limited investments but with a wide array of benefits.

After all, innovation in its most basic form is any new idea, method or product that results in an improvement.

Innovation is possible in almost anything. Even trivial and ridiculous processes can be transformed using simple techniques and methods. This makes it possible for engineers engaged in mundane tasks to awaken their creativity.

During my years as an engineer, I came across many other engineers who were assigned various tasks, and found ways to either simplify them or automate a part of it, thereby speeding up the process. None of these efforts would normally be considered as innovations by our rather lofty expectations of the word.

However, these were indeed examples of innovation, and in almost every case, the innovator emerged as a more accomplished engineer.

Social media is full of posts exhorting engineers to find hidden challenges in their jobs and to let their creativity go wild in even the most mundane tasks. Nevertheless, most of these advisory posts do not consider the undercurrents that exist in any company. They assume every innovative effort by an engineer will be appreciated and rewarded by a company.

As described in the previous chapters, management will, for various reasons, find ways to put down talented engineers. The two reasons described in the previous chapters was that management pursues mediocracy as the baseline or attempts to harass an employee into submission. The pursuit of mediocracy results in talented engineers being perceived as a threat, which in turn, results in them being sidelined. Harassment, on the other hand, attempts to break the spirit of the engineer, in which innovativeness is belittled or criticized.

This may sound ridiculous to a rational thinker, but for anyone who has spent time in industry, this is a sad story many either experienced or witnessed. Those who experience it end up being extremely bitter and rightly so, as instead of being rewarded for enthusiasm, they are punished.

The more talented an engineer, the more devastating the effect, as one begins to feel one's skills and talent are a burden that one has to bear—instead of an asset. Some will lose their enthusiasm and disconnect from their jobs. Others will leave their companies and hope new jobs will treat them better, only to find that most companies follow these nefarious practices.

The above arguments might lead the reader to believe

innovativeness is a curse rather than a blessing. However, a double-edged sword can hurt the bearer as well as allow the bearer to hurt others.

Thus, as Machiavelli said: *it is the armed prophets who have succeeded, and along with bearing arms, the prophet needs to be proficient in using them.*

The same can be said about innovativeness. A company rife with nepotism and harassment is more likely to perceive an innovative employee as a threat that needs to be cut down. Therefore, for a talented engineer enthusiastically trying to find ways to improve processes, it is advisable to first examine the nature of the company, as not all companies will encourage such enthusiasm.

From the discussions in the previous chapter, the inherent nature of business results in most companies being profit-driven, and unless there are mechanisms deliberately created and maintained by leaders with vision, the narrow-mindedness of most managers will result in companies where innovativeness either has little value or is seen as a hindrance.

With all this said, *how then does one awaken one's innovativeness and put it to good use?* I return to the statement made in the previous chapters, that a job is an exchange of services in return for compensation.

A job in a company is not and should not be what defines an engineer.

The rare exceptions to these are those few companies that are very progressive and continuously strive to create an environment where talent can grow. Such companies are few, and it would take a visionary leader to ensure such an environment prevails, as there are always those, who for various reasons, would like to see such freedoms extinguished.

In many situations, it is advisable for engineers to develop a strong professional identity independent of their companies, as their companies might either have no appreciation for innovativeness or may find a way to hurt the engineer.

The Machiavellian Engineer

There are many options for engineers to carve out an identity independent to their companies. The simplest option is to revisit old projects they might have undertaken before they began working in their present company. The best projects to rekindle are those that were conducted as students, as intellectual property that has been acquired in academia is usually the least stringently protected.

Every engineer will undertake projects at various stages during their academic programs. The higher the degree of qualification completed by an engineer, the greater the exposure to projects and ideas. The holder of a doctorate would definitely have a greater resource of past projects than someone with a Bachelor's degree.

However, it is important to stress, that any idea can lead to a fruitful enterprise, and one does not need an advanced degree to pursue a novel idea.

As before, it must be emphasized that a project need not result in ground-breaking discoveries for it to be worth pursuing as a side-project. Sadly, most engineers look back at their past projects and eliminate them as insignificant and unworthy of restarting. It should be noted that a side-project does not warrant the same feasibility analysis as a regular project. When undertaking a regular project, an engineer would typically examine competing products or services, the current and future demand for the product or service, which would eventually justify whether the project was worth undertaking.

Using this same feasibility analysis in starting or restarting a side-project will for the most result in that engineer being bogged down by all kinds of uncertainties, eventually making it impossible to find the enthusiasm to start the project. This distinction between a regular project and a side-project is extremely important, and therefore will be examined in greater detail.

In a conventional project, one needs to perform a feasibility analysis to get approval for the project. This is essential whether one works in a company or is trying to start a company, as every project needs funding, and those who approve the financing of a project will examine the feasibility of the project plan.

This is the case even if a project does not need physical resources such as materials and equipment, as in a company, the cost of the project includes the hours spent by engineers working on it, may it be on gathering information, design or simulation.

A side-project on the other hand, does not need such accounting, and if the physical resources needed are minimal or limited, the only major resource is the engineer's time and energy. Therefore, side-projects can be started incurring minimal cost and therefore, minimum risk. One might argue that to bring an engineering project to fruition will eventually require significant physical resources. However, these costs could be deferred to a later stage when the project has advanced to a great extent.

With this said, *what should an engineer consider when starting a side-project?*

Here, I would like to recount some of my experiences and cite a few other notable examples. For the past decade, while working in several companies, I pursued my side project which has been the development and maintenance of a circuit simulator for electrical engineers.

I started this project for several reasons.

As an electrical engineer and researcher, simulating circuits was something I needed to do regularly. There were numerous circuit simulators already available, and I had used many of them. However, almost every simulator I used had shortcomings. I enjoyed simulating complicated circuits, and the process of simulation has fascinated me.

Circuit simulation was not a novel concept and therefore, building a circuit simulator would by no means be a ground-breaking innovation, and often I am asked by many others as to why I chose to build another circuit simulator when many others already existed.

If I considered this one of the main factors behind starting a side-project related to circuit simulation, I would never have started the project. All that mattered was that every electrical engineer used circuit simulators and the nature of these simulators varied vastly according to the specific application. No single circuit simulator could be proposed as a universal circuit simulator that could be used for every application, though there were many that were popular in particular domains.

This implied that any new circuit simulator could be constructed with features that would serve a given set of applications. In general, when choosing a side-project, an application with high demand and a great deal of variability is a good choice, as one can always find reasons to justify the project.

Another reason for a side-project is to create something that one regularly uses. In the case of my circuit simulator, this was one of the driving forces behind my creating my own application—to have an application that I have full control of.

There are many notable examples of projects, particularly in the software industry, where developers created applications to fulfil their own needs. A few popular examples are of the well-known Linus Torvalds who created the Linux kernel as well as the distributed version control system Git, and Guido van Rossum who created the programming language Python. These projects inspired me to create my own application.

As a computer scientist, Linus Torvalds used the Unix operating system. Though Unix was a favorite of many programmers, it was not used by the wider world due to the difficulty in configuring the operating system. He thought of an operating system built around a kernel similar to Unix, but which would be much easier to install and configure.

He began the project after creating a rudimentary version, sent out an email to several other programmers, asking them to try out his operating system, while also providing them with the source code. Interest quickly grew with many not only using his operating system, but also contributing code to it.

The very first version of Linux was released publicly in 1994. Today, Linux powers most of the servers in the world besides also being the foundation of Android, which runs most of the smartphones. During the early years of Linux, the mainstream operating system providers rubbished the project, claiming it would amount to little more than a hobby for a few software enthusiasts. However, now, anyone who holds a smart phone in their hand is holding a device that runs on an operating system based on Linux.

Surprisingly and inspiringly, Linux is not the only creation attributed to Linus Torvalds. He also went on to create Git, which is a version control system that almost every software developer uses and is required to be familiar with.

An interesting episode led to the creation of Git.

When software developers collaborate on a project, they use a version control system which allows them to track changes in the programs and associated files that comprise the software. The Linux project was using a proprietary version control system called BitKeeper. In 2002, the copyright holder of BitKeeper withdrew the free use of the software. Not satisfied with the other version control system applications freely available at that time, Linus Torvalds created Git to match his requirements, and once again released it open source.

Guido van Rossum created the Python programming language in 1989 out of boredom. During the Christmas of 1989, he found himself with nothing to do, and instead of simply idling away his vacation watching movies, decided to find a "hobby programming project that would keep him occupied during the week around Christmas."

He built an interpreter for a new scripting language which had been on his mind for a while. He chose the name of this interpreter to be Python—a name he chose being a fan of the popular comedy troupe Monty Python's Flying Circus.

Since then, Python grew tremendously with applications in numerous domains. Today, the Python programming language is one of the most popular programming languages in the world used by some of the biggest companies. Guido van Rossum was named as the *"the benevolent dictator for life"* for the Python project, though he decided to step down from the role in 2018.

My reason for citing these two examples was to describe the motivation for launching projects. *Could either Linus Torvalds or Guido van Rossum have guessed that what they started would end up transforming the world?*

I doubt it.

These were examples of extremely talented people who launched projects either because they themselves felt the need for an application, or simply needed an outlet for their creativity. Neither had a business plan or undertook a feasibility study to examine how their application would be better than what was already available. Besides the two examples, numerous scientific advancements happened because scientists tinkered around without a plan or objective.

To sum up my assertion that talented engineers should seek challenging side-projects independent from their companies, I put forth this quote from Machiavelli:

> *Let him act like the clever archers who, designing to hit the mark which yet appears too far distant, and knowing the limits to which the strength of their bow attains, take aim much higher than the mark, not to reach by their strength or arrow to so great a height, but to be able with the aid of so high an aim to hit the mark they wish to reach.*

For any talented engineer, the greatest reward is to be accomplished and acknowledged. Though not all possess the skills needed to gain worldwide renown, one should strive to emulate the great ones who did so.

This was my hope when I launched my project of creating a circuit simulator, though knowing fully well that it would never be a renowned project, let alone have a transformative effect on the tech world. But as they say, hope is a wonderful thing, and if hope can at least make our lives a little less dreary, then that hope is wonderful indeed.

Let us now move on to the question of how one should start a side-project. I cannot emphasize enough on how one should never start a side-project in the same manner as a regular project.

The reason being: a side-project is undertaken for one's own satisfaction while a regular project is undertaken for remuneration.

For any engineer, to find the motivation to launch and sustain a side-project will be difficult, and even more the case when one finds oneself in a toxic workplace. Obstacles will be many, goals too distant and abandoning a project that is unnecessary for one's sustenance will be natural. Therefore, when starting a side-project, keep in mind the fundamental reason for the project, namely an outlet for one's creativity and a safe space where an engineer can work without any interference.

The very first steps in any side-project, as with any undertaking, will be the most difficult. It is advisable to choose at the very beginning of the project, only those tasks that are enjoyable. Unless one is unusually determined and motivated, minute planning and detailed organization will make a side-project agonizing. When one is already exhausted, both physically and mentally at the end of the workday, unless an activity is enjoyable, the urge to put it off and instead relax is too great.

Therefore, in the very beginning of a side-project, clearly distinguish between the tasks that are enjoyable versus those that are not, and focus on those that one would genuinely like to indulge in.

For any project planner, this is anathema. As already stated, this is the special case of the embryonic stage of a side-project, and is inadvisable throughout the duration of the project.

It can take a few weeks or months, depending on the nature of the project, before an engineer is genuinely interested in it. These formative weeks are important in the survival of a side-project when the engineer bonds with the project.

As someone who undertook numerous side-projects but only saw a few survive, I can attest to how the survival of the project was related to how I enjoyed the early stages of the project, and began to view it as my creation.

For those projects where my initial enjoyment was lukewarm, my interest waned quickly and I eventually abandoned them.

Once the engineer overcomes this first bonding stage with the side-project, the project transforms from a fun activity into something one must indulge. At this stage, the project gradually replaces hobbies one indulges in one's leisure, such as watching television or playing video games.

When this stage arrives, the side-project is a serious venture. Unless there are other adverse circumstances such as poor health or other personal problems that could divert the attention of the engineer and cause interest in the project to wane, the project no longer needs special treatment to survive.

At this stage, the engineer can implement rudimentary planning processes to convert it into a fruitful activity. It is important to note that after crossing the initial threshold of project survival, it is still not advisable to resort to conventional project planning techniques, as there are still a few challenges that remain.

The very first challenge is of time and energy.

To sustain a side-project and carry it forward to an advanced stage, one must invest a significant amount of time and energy that would typically be set aside for one's own family and friends.

This is a major obstacle for anyone, as very few wish to sacrifice their personal lives for an additional project that might amount to nothing.

At this point, I will repeat some of the discussions in the previous chapters, as some of the nefarious management practices will be obstacles for any creator. Simultaneously, the solutions presented in the previous chapters will also be applicable in this chapter and a creator can adapt oneself to various work environments to continue working on side-projects.

The natural tendency for any engineer in an exciting project is to discuss it with others, as most engineers seek acknowledgement.

However, one must be cautious about who one brings into confidence in the early stages of a side-project. It is advisable to hide the side-project to anyone in the company. I confess that I am not a legal expert, and in matters where a reader finds the need to examine the finer aspects of their work contracts, it is advisable to get legal help.

This said, labor laws vary greatly across the world, and can also vary across jurisdictions in a country. Beyond what is legal, there are elements in employment contracts, which even if they can be challenged, might be complicated for an engineer to interpret. Therefore, in all cases of ambiguity, seek the counsel of a legal expert.

Let me cite a few examples of employment contracts that are blatantly hostile to any innovativeness on the part of the employee. As someone who has worked in different countries, I worked in certain companies where any work besides that assigned by the company was strictly forbidden.

I have also worked in companies which claimed all work done by an employee as their intellectual property. This also included any work done by the employee on their own time using their own resources.

I also worked in companies where the contract stated that if the employee made any changes to prior inventions while working in the company, the modified inventions would now be the property of the company, and therefore, the company had the right to claim any side-project the employee started before joining the company.

In almost all cases, these clauses in the contract are hidden or are supplied as addendums an employee needs to sign upon joining, giving little opportunity to renegotiate.

I have always been shocked by this brazen nature of companies treating their employees as slaves rather than service providers.

How much these contract clauses were legal and enforceable is something I never took the time to find out, and my attempts to research this topic yielded very different results, as law varies differently depending not only on the country, but also state or province.

However, it was clear to me companies that had such clauses in their contracts were companies that would suppress any creativity on the part of the employee. Legal advice varies greatly, and even articles written by experts in intellectual property will usually leave a disclaimer that expert advice should be sought in specific cases.

In any case, the threat posed by companies taking such a posture to the independent innovativeness of employees should make an employee cautious about disclosing any side-project to any other employee in the company.

The reason for the above digression was a word of caution to anyone who indulges in a side-project. Unless the company has an open policy toward employees undertaking side-projects, and there are clear past cases where such side-projects were encouraged, not punished, it is advisable for the engineer to separate the side-project completely from the company.

Though seeking legal help might produce different results, frankly, a legal battle with a company, particularly a large and well-established one, is unappealing to any engineer. In these modern times, companies reserve the right to hire and fire at will, which implies an engineer can be terminated if the management has reason to believe that they are engaged in a project independent from the company.

Though labor laws might prohibit unjust termination, a legal battle against termination should be the last recourse.

When a side-project crosses this first hurdle and it is no longer necessary to find the time and energy to work on it, but rather the engineer views the project as a haven and seeks comfort and pleasure from indulging in it, the project becomes sustainable.

As already stated, this hurdle is a major one, and is a significant achievement. I would like to take a moment to describe some of the aspects of this stage, as only those who have overcome this hurdle can truly speak about it. During the early stages of a side-project, the motivation for an engineer is the novelty of the project, which for several reasons may be enjoyable. This could be due to the opportunity to perform tasks which one could not in their official jobs, or a feeling of nostalgia when one rekindles an old project.

Gradually, as the weeks pass by, the interest in the project gradually grows almost similar to a romantic relationship, until the side-project becomes the principal interest of the engineer. This transformation is almost magical and one will wait for the work day to be done to return to the side-project.

Overcoming this hurdle also coincides with a transformation in the engineer.

With an engaging side-project one is interested in, the dreary and mundane nature of work is no longer agony. On the contrary, not having to mentally challenge oneself at work, which would normally have been a source of frustration, will now be a blessing, as the engineer can now channel their energies to the side-project instead.

Such diversion has the advantage that the side-project is no longer controlled or influenced by the company management, but is driven and regulated solely by the engineer. As a result, the side-project becomes a place of refuge when an engineer can feel truly engaged and free from the toxicity of the workplace.

Over the years I spent in industry, my side-projects kept me insulated from office politics and favoritism, as the pleasure I derived from my side-projects made my jobs nearly irrelevant.

In companies that adopt favoritism and harassment as policies against talented employees, a side-project can be the best antidote for the employee. The engineer must put up an appearance of submissiveness and obedience in the company to pacify management, while the true talents of the engineer are being utilized in side-projects.

Interestingly, favoritism and harassment that might break an engineer will turn into a joke as one whiles away time in the office and laughs away the silly politics plaguing most companies.

As the effort and interest in the job wanes, it might happen that the talented engineer will gradually slip into the ranks of the mediocre. While the engineer's skills and capability remain high, what is invested in the company is limited, and is no longer perceived as a threat.

Some of this might seem ridiculous, but as someone who spent time in many different companies, I attest to this occurring in most of the companies I worked for. By finding a truly interesting objective, many problems were simultaneously solved.

The essence of the solution lies in understanding one's options when one is subservient without power. Those in power wish to control the powerless, but, except in the most extreme circumstances, no one can control another individual completely, least of all their thoughts.

Though it is possible to punish someone for actions and speech, it is almost impossible to punish someone for one's private thoughts.

A side-project unlike a regular project need not have a conclusion, nor even for that matter have the stages that a regular project is expected to have. I spoke about the early stages of a side-project, and how it can transform the mentality of an engineer.

Let us now talk about the options available to an engineer once he or she is engrossed by a side-project. This depends on many factors, professional and personal. I cited a few examples of side-projects that transformed the world.

But not all projects need to be so.

On the contrary, at certain times, attempting to take a side-project to such milestones, can be a burden. At this stage, human nature and the personality of the engineer plays a significant role, which is something Machiavelli spoke about and which will be discussed in a later chapter.

Let's examine the simplest and safest option.

An engineer pursuing a side-project decides to not expand it, but to leave it as that enjoyable activity that engrosses and offers a place of refuge. This might seem like a waste of skill and talent any book on entrepreneurship would caution against. Pointedly, for some engineers, there is no reason to seek renown. An engineer may consider family to be important, thereby limiting the time available to invest in the side-project. This is by no means a fault, as wishing a happy family life is not a selfish and unreasonable desire. An engineer might already be burdened by debt, in which case, the risk associated with expanding a side-project, may not be acceptable. An engineer working in a large organization, receiving good remuneration and benefits may not want to jeopardize it, and therefore, might want to limit the side-project as a coping mechanism, to deal with a workplace that has no appreciation for skill and talent.

The next option is to pursue it as a scholastic venture rather than one that would be a mainstream occupation. In such a case, a side-project may generate either minimal or no revenue but will serve to project the creator or creators of the side-project as experts in the domain. Such an option is usually favorable for an engineer working in a large corporation, where having an acknowledged expert as an employee might be beneficial to the company. The company may not object to, if not actively encourage, an employee spending time and energy on such a project.

Moreover, if the project is labeled as a scholastic venture, with no monetary benefits, it may be deemed a minimal threat to the business of the company. Most notable examples are open-source software created by engineers while working in large corporations. Even in the absence of monetary benefits, the acclaim and acknowledgement that engineers receive, results in greater employability and freedom of movement within the industry.

The last option is to convert the side-project into a mainstream project that brings with it not only acclaim, but also monetary benefits. This option now steps into the realm of entrepreneurship, and therefore, will not be discussed in detail in this book, as the author is not an entrepreneur and can offer no useful advice in this regard.

It should be noted that transitioning from an engineer to an entrepreneur requires a significant transformation of perspective and approach. While demonstrated successfully by some, for most engineers, this is daunting. If one wishes to take the plunge, research those who succeeded. There are many biographies to study.

For those who choose the first two options, let's elaborate on several aspects that can improve the success rate of a side-project. As a side-project grows and it becomes clear to an engineer the project has potential, it is only natural to seek ways to increase the time and energy invested in the project.

Depending on the nature of the engineer's employer, there are several possibilities. As already stated, there are many different types of companies—mega corporations, family businesses, start-ups, mid-sized corporations and many others. This was discussed in the second chapter, and in this chapter, some of the intrinsic features of business types can be used to devise strategies to expand on one's side-project.

In the case of start-ups and small businesses, one finds the least risk in expanding a side-project. To begin with, small companies do not employ extensive resources on scrutinizing the behavior of employees, for the simple reason that funds and personnel are limited. Therefore, engineers working in a small company find themselves with a significant amount of freedom on how they can use their time and energy.

Moreover, small companies are driven more towards results and successes rather than process. This means, as long as an employee delivers on basic goals and objectives, they have a good deal of excess bandwidth to indulge in their side-projects.

One usually equates a start-up or a small company with long working hours and high pressure. However, as previously cited, in most cases, the long hours are a show of presentism, and if one spends significant time in the office and appears to be busy, one is rarely questioned about the details of the work done.

Thus, by reference to the prior chapters, the tactic of salami-slicing can be effectively employed to divert time and energy to one's side-project, even though this time and energy should have been devoted to the company.

As the company grows larger, the challenges in sustaining a side-project increase. Large corporations have significant resources at their disposal to secure their workplace. Computers and laptops given to employees have spyware to monitor how they are being used. One cannot access public websites or open one's personal email. One cannot connect a removable storage device except for those that have been provided by the corporation itself.

These factors limit how the employee can spend one's time in the workplace, and therefore, make it more difficult to sustain a side-project. Moreover, larger corporations are more process-driven, which implies a greater amount of time spent in meetings, group events and interactions, so employees are restrained from working or behaving as they please.

In such cases, careful planning is needed to detect patterns, to be able to know times when one could be relatively free, as opposed to times when one might need to be diligently working.

One might be disturbed by these observations and think one is expected to work diligently throughout the official work period. However, in these modern times, the official work period has either been extended or completely left open, with employees expected to be always available.

Part of this is justified, as in this new era of globalization, one finds oneself working with people all over the world, or serving clients in different parts of the world. Therefore, the erstwhile 9-to-5 jobs have now shrunk to a few service sector jobs, with most other professionals expected to be available at odd hours.

If the older principles of work ethics are used, one would never find time for oneself, let alone time to pursue a side-project. In these modern times, one must differentiate between genuine working time and the time for which one needs to be available.

As mentioned in the previous chapters, the intrinsic nature of business which requires a company to be profit-driven, forces a company towards repeatability. Therefore, in most companies, a significant proportion of the tasks performed by employees will be repetitive and mundane. These are essential to delivering products and services of a quality sufficient to ensure the company remains profitable.

Beyond these tasks, companies expect their employees to remain committed and ensure that they put in a good effort.

Unfortunately, human nature tends to reward those who spend long hours in the workplace, interpreting this as showing the greater level of commitment. All one needs to do is to spend those long hours in the office to please the management, while surreptitiously diverting a portion of official working time for one's own activities such as side-projects.

It might appear deceitful, but one has to remember this world is not governed by fairness and equity, to which I would offer this quote from Machiavelli, which may come forth as a bit of a repetition, though it is worth examining again:

> It is necessary, therefore, if we desire to discuss this matter thoroughly, to inquire whether these innovators can rely on themselves or have to depend on others: that is to say, whether, to consummate their enterprise, have they to use prayers or can

they use force? In the first instance they always succeed badly,
and never compass anything; but when they can rely on
themselves and use force, then they are rarely endangered. Hence
it is that all armed prophets have conquered, and the unarmed
ones have been destroyed.

As an employee, one will find one's work continuously increasing, while one's remuneration and benefits do not match. Therefore, if one plays by the book, one plays to a disadvantage, as the book is usually biased pitted against the employee and in favor of the management.

An employee should ensure their activities are not a flagrant violation of company policies and does not hurt the business of the company or appear as a conflict of interest, where the side-project could threaten some aspect of the company's products or services.

Employees usually fear that their side-projects could result in reprisals from the management and threaten their regular jobs.

However, one should examine the nature of the company and prior cases. Often, it is advisable to not disclose a side-project to the company, especially if one is not deriving monetary benefit from it.

If there is reason to believe a company will have a positive view of an employee pursuing a side-project, disclosure can be better to ensure an environment of trust. Such might be the case, when companies explicitly state in their work contracts that any work done by an employee outside the domain of the work of the company, will remain the property of the employee.

An employee might fear that the company management will distrust them if they pursue side-projects. To this, my answer is that a talented employee should not seek to appease management, as that would lower them to the level of the flatterers who seek to retain their position by pandering to shallow and petty managers.

A talented employee need only deliver on assigned tasks

with diligence to retain their positions in the company. As earlier indicated, in most companies, the level of effort is not too challenging for a talented engineer. Beyond retaining their positions in the company, talented engineers should seek to expand on their skills and knowledge. This chapter already described in detail the benefits of side-projects.

To conclude this chapter, I emphasize that—except in a few rare cases when engineers find themselves in progressives companies with visionary management—they will not be assigned challenging or innovative tasks, nor will their innovativeness be encouraged or rewarded, and worse, might find themselves in a hostile environment that will destroy their spirit.

Later chapters will describe aspects of human nature that result in such behavior. However, the purpose of this chapter was to describe how following an independent side-project can provide a wonderful refuge for a talented engineer where they are free from whatever mess their company may be.

Moreover, a side-project can drastically change the outlook of an engineer and improve their mental well-being—despite dismal and mundane working conditions.

As Machiavelli advised: anyone choosing a new path can expect to face a range of hostilities and must be willing to be armed and combative. For an engineer pursuing a side-project, it is essential to be on guard against the company choosing predatory practices to either scuttle the project or cut down the employee.

In such cases, the employee should adopt a range of deceitful measures to keep management at bay. The greater the engineer can rely on their own abilities and efforts, the greater are the chances the project will be successful.

Beyond Ability

THIS CHAPTER AND the chapter *Of Human Nature* were difficult for me to separate when comparing Machiavelli's writings. However, I decided to divide these two chapters on the basis that the chapter *Beyond Ability* is descriptive of actions that a prince might be compelled to undertake, while the chapter *Of Human Nature* describes the personality a prince must project to successfully rule.

In this chapter, I bring together three separate chapters of Machiavelli's writings—*Concerning New Principalities which are either Acquired by the Arms of Others or by Good Fortune*, *Concerning those who have Acquired a Principality by Wickedness*, and *Concerning Cruelty and Clemency, and Whether it is Better to be Loved than Feared.*

The reason for this combination is that in all these chapters, Machiavelli describes how a ruler might be forced to act in a shocking and contemptible manner, but such actions—if executed properly—can result in success.

The depth and variety in Machiavelli's philosophy is truly remarkable given the fact that he analyzed so many different events and attempted to find reasons for different outcomes.

Machiavelli acknowledged even five hundred years prior to

his time, there was no one-size-fits-all approach to governance. There were rulers who did all that they should have and were still ruined, while others succeeded with those same strategies.

Therefore, in cases where a ruler failed, there were other factors that may not have been obvious, and in some cases, the circumstances were such that extraordinary skills and capability were required. In a separate chapter on how a ruler can withstand the vagaries of fortune, Machiavelli also acknowledged the impact of luck.

The same can be said of the corporate world, where some iconic brands failed, though at the surface, all that could have been done to prevent it, had been done. Many are quick to blame corporate failure on luck, but often, more complex reasons exist for the downfall.

During my years in academia as a graduate student and later in industry, I closely observed many professors and managers. In all cases, I could correlate their intrinsic personality and actions, and subsequently observe whether they succeeded or failed.

In a manner similar to Machiavelli, one could discern red flags in the management of a company an engineer should be wary of.

Machiavelli speaks of the case of when a ruler ascends to the throne by either good fortune or is installed by another power. However, this excludes the case of a dynastic rule when a member of the ruling family ascends to power, as in such cases, it was an expected event for which many provisions had already been made to ensure that the ascension was smooth and successful.

The case spoken of by Machiavelli was usually when the heir to the throne was someone unexpected or unprepared. It could be that a captain or commander in the conquering army was made the prince or a private citizen rose to the throne with the support of the invader who wished to establish an oligarchy, but decided to execute all members of the erstwhile ruling family to eliminate any potential for future rebellion.

Under such circumstances, unless the new ruler had great

ability, they could rule only at the pleasure of those who had installed them on the throne.

> ...he who has not first laid his foundations may be able with great ability to lay them afterwards, but they will be laid with trouble to the architect and danger to the building.

To ensure the new ruler could rule as he pleases, he must consolidate his power after ascension which as Machiavelli states, is difficult and often risky, as it needs a rebuilding of the power structures around him.

This is often the case.

Those who installed the new ruler and those who offer their support may have ulterior motives. In such cases, it is inevitable that the new ruler will resort to acts of cruelty to rearrange this balance of power in his favor by dispossessing some and rewarding others.

Machiavelli gives a detailed account of how Duke Valentino attempted to retain power after the death of his father Alexander the Sixth, and even though most of his actions were commended by Machiavelli, he eventually failed as he was himself on the point of death.

The history of Cesare Borgia, who later came to be known as Duke Valentino, features in a great deal in Machiavelli's works, as Machiavelli was the accredited agent of the Florentine Republic governed by Cesare Borgia and examined the Duke very closely.

The Pope Alexander the Sixth, who also featured in Machiavelli's writings a great deal, wished to launch a conquest of Italy, but was deterred by the many factions that existed at that time. The only solution to breaking the political deadlock was to bring an external power, namely the King of France.

This was achieved by the Pope promising to annul the King of France's marriage in exchange for a military alliance. The Pope was instrumental in bringing his illegitimate son, the Duke Valentino, to the forefront with the Duke leading many of the military campaigns in Italy. This is where many of the discussions featured the specific qualities of the Duke, as according to Machiavelli, the Duke performed admirably well, given the turmoil of those times.

Duke Valentino possessed almost every virtue—an intrepid military commander, a crafty statesman, and a ruthless murderer; therefore, according to Machiavelli, the Duke did everything that could be done, when a prince acquires a principality which was bestowed on him.

On the military front, the Duke led many successful military campaigns, initially using the armies of allies, but then created his own armies, after disbanding those armies whose loyalty he found questionable.

On the diplomatic front, he used money and promises to win over petty princes in Italy, and to turn the Church in his favor. He brutally murdered those who could not be bought but might have stood in his way. Machiavelli stated his only error was to not oppose the election of the next Pope as one who was not his loyalist. However, in the end, the campaign ended, with not only Alexander the Sixth dying, but his son the Duke, also at the point of death.

The detailed actions of Duke Valentino are interesting, and though written in a manner that is difficult to understand unless one reads the chapter several times, there is an indication of the challenges faced by someone who ascends to the throne without prior preparation:

> *Therefore, he who considers it necessary to secure himself in his new principality, to win friends, to overcome either by force or fraud, to make himself beloved and feared by the people, to be followed and revered by the soldiers, to exterminate those who have power or reason to hurt him, to change the old order of things for new, to be severe and gracious, magnanimous and liberal, to destroy a disloyal soldiery and to create new, to maintain friendship with kings and princes in such a way that they must help him with zeal and offend with caution, cannot*

find a more lively example than the actions of this man.

The above concluding statement by Machiavelli on Duke Valentino would have anyone trembling at the thought of what a ruler needs to do hold his kingdom.

However, this is merely an antidote to a fickle and cruel world, where even among humans, the law of the jungle prevails.

In all my years, I saw many a talented engineer flounder because they were unable to handle the intentions of those around them. Except for a few rare cases when talented engineers received adequate support and reward, in most cases, the talented were either neglected or worse: suppressed.

In addition to skills and talents, one must possess the other qualities mentioned above to effectively see one's hard work bear fruit.

What is striking in almost all the examples presented by Machiavelli is the disturbing fact that violence was a component.

Machiavelli considered this violence as a prerequisite for statecraft—without which power cannot be rearranged. Violence was a way to subjugate people, instill fear in them to prevent them from rebelling in the future, and a way to redistribute the limited resources within a kingdom.

A ruler would dispossess a ruling family and execute the heads of the family to transfer the family's property to another, who the ruler had wished to gain as a friend.

In doing so, the powerful, future threat was rendered powerless and unable to cause harm, while a friend was bestowed with benefits. This ally would be loyal and carry out the bloody tasks of the ruler in exchange for protection from rivals or other potential invaders.

> *Such stand simply elevated upon the goodwill and the fortune of him who has elevated them—two most inconstant and unstable things. Neither have they the knowledge requisite for the position; because, unless they are men of great worth and ability,*

it is not reasonable to expect that they should know how to command, having always lived in a private condition; besides, they cannot hold it because they have not forces which they can keep friendly and faithful.

During the medieval times that formed the basis of Machiavelli's writings, princes were of different types. There were those who were powerful and ambitious, starting wars and leading military campaigns. There were those who survived by allying themselves with a larger power and changed their allegiances according to necessity; some being successful and remaining satraps to a larger power, while others were destroyed by larger powers who found them untrustworthy.

There were also a few with ambition, but lacked the ability to lead a successful military conquest and were ruined. Unlike modern times when an unsuccessful politician accepts retirement without ascension to power, in those medieval times, an unsuccessful prince would usually be executed.

In the modern corporate world, except in rare cases, one does not have to contend with the same level of violence as would be present in a war. Therefore, an unsuccessful manager will rarely see their career come to a complete halt unless some other factors come into play. During my years in industry, I observed several managers who were unable to retain their positions, either leading them to look for positions external to the company, or choosing to return to a technical position.

Also, there were many other managers barely retaining positions—clinging by their fingernails—in cases similar to petty rulers content with being the oligarch of a more powerful ruler and content with keeping upper levels of management satisfied enough to not be replaced.

As an engineer, to find oneself being led by a capable manager is rare, and a vast majority of engineers will find themselves with managers who either struggle to retain their positions by various means or who themselves move around between companies as they find themselves unable to succeed in one company. Working under managers who themselves are struggling to retain their positions is an unsatisfying professional experience as these managers indulge in an array of nefarious practices to survive, such as harassment, favoritism, factionalism and many more.

For an engineer, there is very little one can do when one finds oneself with such a manager. Officially, there is the possibility of approaching human resources, this usually provides little relief to an engineer as the human resource department is typically tasked with protecting the company. Unless the manager indulges in practices that actively hurt the company's business interests, rarely will any appropriate action be taken.

The actions of the manager, of course, directly impact the well-being of an engineer. In an earlier chapter, I described how managers use favoritism and choose mediocrity to retain their control over their teams. The violence princes resorted to in medieval times to control their nobles and subjects, however, is replaced by several other strategies, such as harassment, neglect, and obstacles to one's professional success and progress, in addition to favoritism as discussed here before.

These nefarious practices cannot be ascribed to the nasty and petty nature of managers and colleagues, but in many cases, is an active, conscious strategy used by management. In this chapter, I use Machiavelli's writings to outline how he described the actions of medieval princes. Strategies should be categorized as effective or ineffective, rather than good or evil.

To understand the nature of one's manager, every engineer needs to meticulously review their manager and evaluate every action and every decision taken by the manager. Not with the objective of offering feedback to the manager, of course, but rather to quantify the capability of the manager as a leader.

A manager's capability can be assessed with respect to their day-to-day activities, dealings with people under normal circumstances, approach to crisis management and behavior to people in times of crisis. Here, parallels are drawn to how Machiavelli believes a ruler should behave.

To begin, since Machiavelli writes primarily about military conquests due to the turmoil present at his time, let's begin with analyzing and understanding Crisis Management.

Engineers must first acknowledge that the manager is just a human being whose primary objective is their own prosperity and security. Unconditional love as if from a parent should not be expected from managers or professors who are responsible to their higher ups, and thus need to do what is needed for their team's performance to be at least satisfactory.

Hereof, never ever expect your manager or professor to be an angel. Such expectations are the beginning of workplace frustrations. The next chapter will describe how an engineer should transition to a managerial position, the differences in the roles and responsibilities that can be expected in such a transition, and describes the differences between an engineer and a manager.

When a manager is under pressure or aspires to achieve something out of the ordinary, this is a simple case of Crisis Management. Crisis Management does not have to be in extreme cases when the company is about to be dissolved or forcibly acquired. Any event other than simple day-to-day management can be viewed as crisis management.

Depending on the domain the company operates in, the size of the company and the nature of competitors, crisis situations can occur either frequently or rarely. In smaller companies or start-ups, almost every day can be a crisis. In mega companies, times of crisis might be at the end of a quarter or the end of the year when reviews or appraisals occur.

In a time of crisis, the very first expectation from a leader is timely detection. As Machiavelli states, *a ruler who resides in a principality is able to perceive troubles and disturbances as they arise.* For a manager or a professor to perceive a crisis before it takes hold, it is extremely important that they are closely engaged with their team. Even though they may be closely engaged with the team, unlike a ruler who needs to perceive military disturbances, a manager needs to perceive disturbances that are far subtler. These disturbances could be underperforming or failing projects, employees who underachieve or act out of malice, and many other problems such as valuable employees leaving or being poached by a competitor.

Inability to perceive disturbances can take several forms— denial that such a disturbance exists, or assuming that the disturbance will not grow to pose a serious threat, or even worse: complete apathy assuming the final consequences will have no effect.

Except in cases of major disturbances such as earthquakes, war or natural calamities, which are sudden and not under human control, other disturbances are typically slower and being unable to detect them is in most cases inexcusable.

A comparison can be made to a frog dropped in a pot of boiling water. It will immediately jump out. If a frog is placed in a pot of water and the temperature is gradually increased, the frog will die without attempting to escape. Therefore, major disturbances—despite being disastrous—are easier to detect, but minor disturbances though are tougher to detect and if left to run their course can be equally destructive.

As an engineer, be wary if you have a manager who cannot perceive disturbances. In the event of a disturbance running its course and affecting the team, the worst affected will always be the junior team members, which includes the engineer. In this regard, faith in an incompetent manager will cost an engineer dearly. As an engineer, it is important to remember that the manager's job is to ensure the well-being of the team, which in turn fulfils its goals such that the manager can retain their position. Inability to do so is completely inexcusable, and if engineers absolve the manager of this duty, engineers are equally responsible for whatever mess they find themselves in.

Human nature plays a major role in being able to perceive disturbances. A complacent manager will be less likely to look for signs of disturbances and even when these signs are obvious and clear, might look for excuses to disregard them.

In my years of experience, most managers I worked with exhibited toxic complacency to various degrees. The most distressing conclusion I drew from observing complacent managers and professors was that it was an intrinsic quality irrespective of prior background.

As Machiavelli stated *"…men are content when their property or honor are not touched."* Someone who struggled in the past was as likely to be as complacent as someone who had a smooth and comfortable life. It is as if humans would like to forget past hardships and hope that the future is rosy.

This makes it difficult to predict if a manager will detect a disturbance in a timely manner. Even if past behavior is promising, the hope for good times might cause a different behavior in present circumstances. Thus, even if one has historical data leading one to believe a manager is or isn't proactive, one should be circumspect about using this knowledge. Present or future behavior might be caused for many different reasons, unlike past behavior.

These arguments can be applied to an engineer as well. Engineers must also be on the lookout for potential disturbances that might threaten their jobs. The complacent sense a job is secure can be costly.

A crisis for an engineer can either be technical in nature when projects get stalled or are scrapped or can be non-technical when there is a change in management or the team is divided or merged with another. Unfortunately, most engineers turn a blind eye to issues which are not specific to solving technical challenges in projects.

However, such changes can completely derail an engineer's progress in a company or make an enjoyable job annoying and dissatisfying. An engineer may not be able to remedy a crisis even after detection, as power lies with the management and not engineers. Early detection gives an engineer time to plan an exit from a company or adopt alternate survival strategies if the situation deteriorates later.

The crisis management that follows crisis detection is also tied to human nature, both for managers as well as engineers. Social media is full of posts about the ideal manager who leads by example, supports their team, is the first to make sacrifices and the last to take credit. Social media is also full of the ideal engineers who push themselves to their limits without regard for benefits or rewards—just for the satisfaction of seeing a project succeed.

Unfortunately, reality is far from these ideal portrayals. Considering these portrayals to be the norm can end in disaster. In my experiences, managers and engineers who survived crisis situations are those who made unpopular decisions and committed unsavory—if not cruel—acts.

To fully understand the need for these unsavory actions, let's examine Machiavelli's writings on this topic.

Management under normal circumstances will be examined in detail in the chapters *A Climb To Management* and *Technical Weapons*. However, to complete the discussion in this chapter, the greatest challenge that any manager must face when not under pressure, is to not fall prey to complacency.

It is natural for humans to sit back and relax when things are running smoothly. Machiavelli speaks about this in two different chapters. The first, when examining the actions of the Romans in the chapter on mixed principalities, and the second, when examining the impact of luck in the chapter *What Fortune can Effect in Human Affairs and How to Understand Her*.

In the previous chapter, with reference to the Romans, this quote can be repeated:

> ...nor did that ever please the Romans which is forever in the mouths of the wise ones of our time: Let us enjoy the benefits of the time—but rather the benefits of their own valour and prudence, for time drives everything before it, and is able to bring

with it good as well as evil, and evil as well as good.

And from the chapter on fortune, the meaning of this quote from Machiavelli is loud and clear:

> *So it happens with fortune, who shows her power where valour has not prepared to resist her, and thither she turns her forces where she knows that barriers and defenses have not been raised to constrain her.*

To be complacent in good times and not plan for the proverbial rainy day is completely inexcusable. This applies both to management and engineers. Engineers must be wary when during boom times, the management splurges on useless activities, or worse, indulges in stuffing their pockets through fat bonuses or stock buybacks.

Such a management will terminate employees without a second thought when times change for the worse.

Conversely, engineers are equally culpable when during good times, they do not invest in self-improvement, training and invest in augmenting their own assets. When they find their jobs at risk with changing times, they will plead and appeal for mercy. In these modern times pleading is useless. During hard times, management turns into vultures with little care for their employees.

Machiavelli wrote many chapters on how rulers should behave that were considered at that time very controversial and offensive.

However, he wrote about his analysis of historical events and compared rulers who were deemed just and righteous with those who were cunning and cruel.

He had examined in detail the example of rulers such as the King of France, Duke Valentino, Alexander the Sixth who was the Pope and a few Roman emperors. He wrote about some prevailing while others failed. His analysis is an excellent read, though difficult to comprehend at first glance.

Whether it was cruelty, deception or wickedness, the effectiveness of any one or more of these qualities is determined by the way they are used. The discussion is broken up into a number of chapters as stated in the opening paragraph of this chapter. However, there is a common thread.

Machiavelli introduces a chapter *Concerning Those who Have Acquired a Principality by Wickedness* where he examines those few cases where a prince comes into being through acts of sheer wickedness. Machiavelli dedicated a chapter to describing rulers who rose from low and abject positions—the first was Agathocles the Sicilian and the second was Oliveratto da Fermo.

Agathocles, born as the son of potter, would normally never have any hope of becoming a prince. However, blessed with strength and intelligence, he joined the army and rose through the ranks to become the Praetor of Syracuse—the Praetor being a title granted by Rome to someone who was the commander of an army.

Oliveratto found himself abandoned as an orphan, raised by his maternal uncle, and during his younger days fought in the army and eventually ended up serving Duke Valentino described above.

Like Agathocles, Oliveratto applied his body and mind, and rose to be the commander of the army.

Both men had great ambition and were dissatisfied with being mere military commanders, but rather wanted to be princes of their own kingdoms. In both these cases, with no power supporting their ascendancy to the throne, nor being of noble birth that gave them a right to assert the throne, they grabbed what was not theirs by right.

It is clear from that chapter that violence and wickedness were the only options to break through their obstacles. If they had not grabbed the throne by force and deception, they would never have achieved their goals.

In describing the life of Agathocles the Sicilian:

Yet it cannot be called talent to slay fellow-citizens, to deceive friends, to be without faith, without mercy, without religion; such methods may gain empire, but not glory. Still, if the courage of Agathocles in entering into and extricating himself from dangers be considered, together with his greatness of mind in enduring and overcoming hardships, it cannot be seen why he should be esteemed less than the most notable captain.

With the above, Machiavelli insists that despite his ascent which would never be described by historians as glorious, Agathocles was an exceptional man of great capability. As well, this goes back to his description of the Duke Valentino where he describes a ruler who governs during tumultuous times as one with great ability who can use a vast number of strategies to consolidate his rule.

Hence the strategy of Agathocles, after his ascent to the throne, to successfully govern his kingdom, repulse attacks from enemies and was never conspired against by his own citizens, is attributed to his skills in warcraft and statecraft which Machiavelli stated were absolute essentials, without which no ruler could survive for long.

Since this book is not about modern politics, but rather about modern life in corporations and universities, one might think such an example is irrelevant. However, if we reflect, almost all of us witnessed some who have ascended through nefarious practices.

Examples include a student awarded a degree using unfair means or an employee attaining a position using stolen work.

Some flounder while others flourish.

As children, we were told fairy tales concluding with the defeat of the wicked and the victory of the righteous, with the moral being that evil will be punished.

Per contra, in real life, we see many examples of those who commit wrong to not just evade punishment, but in some cases achieve a great deal of success. When examining such cases, one needs to not just examine the nefarious act committed, but also examine the inherent ability of the person. Those who succeed in the long term have ability and may resort to nefarious acts for several reasons.

It is impossible to delve into all the reasons compelling a person to commit a nefarious act they know to be blatantly wrong and immoral. As with the case of Machiavelli's example, one should acknowledge that the world is not entirely fair, and for some, there are obstacles and disadvantages making it impossible to progress without committing a violation.

Almost everyone acknowledges that laws are created by those who wish to remain in power or wish for the status quo to continue to favor them. Unwritten social norms are often the result of the privileged few who expect others to let them continue to be privileged.

Thus, when one finds oneself blocked by unfair rules and norms, one can be frustrated by a lack of progress or one can take well-calculated risks to overcome or bypass obstacles.

Almost everyone who has worked in a company will agree on how employment contracts are framed to empower companies at the expense of employees. Unless a right is guaranteed by law, a company will either stay silent about it, or worse, encroach on it to either completely violate it, or weaken it significantly.

In the past chapter, I gave a few examples on how companies claim rights over intellectual property generated by employees, and how many of them are a flagrant attempt to either stifle individual innovation or grab what should rightfully belong to the employee. The past chapter provided these few examples in the context of innovation and intellectual property. All the same, the encroachment of companies into the rights of employees exists in almost every aspect of employment, and in a few cases, can be extremely agonizing for an employee.

It is impossible to cover the many instances of contractual overreach in this chapter. However, to provide a context to an analogy to Machiavelli's writings, I will examine just a particular case highlighted in recent times.

Today, it is normal for employees to leave a company and join another every few years, as such a change brings not only the opportunity to negotiate better compensation and benefits, but also fresh challenges and an entry into new domains.

Though a few employees report no issues in this migration, in many cases, leaving a company can be messy. One would think management should not object to employees who wish to leave as there could be many reasons for doing so.

Sadly, some companies pursue official and unofficial policies to harass outgoing employees. As someone who exited several companies in the past decade, I saw managers behave in shocking ways.

Most managers see professional migrations as threats to their operations. By professional migrations, I imply that when an employee leaves a company for another, possibly a competitor in the same industry. In contrast, employees leaving for personal reasons such as to accompany a family member, are dealt with less hostility.

A threat a company perceives when an employee announces an exit is that other team members might follow. The exiting employee might have negotiated a higher salary or landed a better position due to experience gained in that company. Strategies adopted by managers include using legal and extra-legal mechanisms to try to make the exit as painful as possible with the objective of sending a message to the rest of the team about the consequences if any of them attempts a similar exit.

Companies expect exiting employees to give a notice period, during which the company can expect that employee to transfer knowledge to colleagues to ensure projects continue smoothly. Most countries have minimum notice periods mandated by law and companies have the right to insist on at least these minimum durations.

However, rarely will a company limit itself to what is required by law. In most cases, employment contracts will require exiting employees to serve far longer notice periods than the legal minimum. Rarely will a candidate at the time of joining a company contest a clause related to the notice period to be served at the time of exit. Clauses related to a notice period can slip through negotiation unchallenged and resurface only when the employee submits their resignation notice.

Options available to employees to withstand harassment from the management is largely dependent on the company's country or jurisdiction. In some countries, exiting employees do not need to produce an exit certificate such as service certificate, and though an employer has the right to ask for references, a poor reference given to an former employee can be legally contested.

In many countries, even in these times, employers demand candidates to produce certificates and other exit documents from prior companies. With these documents, employers hold exiting employees to ransom, and some, unfortunately, end up submitting to every whim and fancy of the management during their notice periods.

How employees deal with management during their notice periods is of course linked with their personalities as well as the personality of the managers.

The purpose of describing the scenario above was to describe how a facility that was founded to allow both employees and companies to transition smoothly, has become a tool for the exact opposite purpose, namely, to thwart the migration of talent between companies. One can attribute this to human ingenuity enabling any policy to be abused, or to the base nature of human beings, who resort to evil whenever possible.

Though a vast majority of exiting employees submit to harassment without much support, there have been a few cases where the employee managed to turn the tables on the management. It should be noted that many of these tactics used by employees are condemned by one and all, and the official narrative is quick to brand these employees as unprofessional and damaging their professional future.

There have been instances where an exiting employee

behaved atrociously during the notice period once it was evident that management was pursuing a policy of harassment. Stories abound of employees spending their notice period playing video games in the office, watching television serials, reading books or indulging in other unacceptable workplace activities.

Milder forms of bad behavior are being tardy, wasting time in the office and disturbing colleagues. Though such behavior is juvenile and immature, they are sometimes are an antidote to equally infantile management.

In such circumstances, the state of affairs continues until complaints emerge about the exiting employee's behavior and management has no choice but to relieve them immediately. There is a risk associated with such behavior, namely, that management will attempt to poison the water by informing the employee's future company of the incidents, thereby jeopardizing the future job.

Another strategy used by exiting employees who fear management will resort to harassment tactics once they put in their notice of resignation, is *ghosting*. The term ghosting originated in the dating world, where after a couple started dating and one of them decides to put it to an end, they disappear or "*ghost*" the other instead of clear and polite communication indicating they wished to end the relationship.

In the corporate world, ghosting occurs when the exiting employee stops showing up to work without resigning. Eventually, management will remove the employee from their records and disable all accounts associated with the employee.

However, there is no transition from the absent employee to other team members and the team must scramble to figure out how to perform the duties the ghosting employee performed. Of course, ghosting is risky. In many countries, it is customary for an employer to ask a new candidate to produce a service certificate from a past employer, or at least ask for references from past managers. Ghosting employees will not have documents from the company they left.

At this point, let's consider Machiavelli's example of a prince acquiring a principality by sheer wickedness. In cases of bad behavior by exiting employees, their actions were either illegal, unprofessional and intentionally harmful to their employers. Though one is quick to blame the employee for possessing poor integrity, in some cases these incidents occurred because of the employee's inherent bad character.

Often these incidents occurred because the employees had ample evidence of the management's approach to dealing with exiting employees. This would be the case when employees saw management harassing employees during their notice period and fearing they would receive the same treatment.

Under such circumstances, the employee had few options. Seeking legal help might not be feasible and submitting to management's unreasonable demands can be insulting and demeaning.

Under these circumstances, revisit the question about what an exiting employee can do. Though there will be those who insist on professionalism in the face of management's harassment, an increasing proportion believe no one should bend over backwards and sacrifice their well-being.

In the years I worked in industry, I witnessed a number of instances of ghosting and other bad behavior. In many instances, the exiting employees though reviled by the management, had the support and even the blessings of the rest of the team who considered the actions of the employee as justified.

Therefore, similar to Machiavelli's examples of wickedness, these employees had little recourse other than to behave in an unprofessional manner when they found their contracts were unfairly pitted against them.

Following their exit from the company, it was clear the future success of these employees was determined by their abilities rather than the guile they showed during their recent past. Some indulged in similar behavior in subsequent companies while neglecting to invest sufficient time and energy in their skills and knowledge and found their careers languishing.

Others made a fresh start, put their past bad experiences behind them and established successful careers. Often one experiences an adrenaline rush after committing a violation. This is especially so when one evades punishment or retaliation.

When this adrenaline rush after bad behavior becomes modus-operandi and the engineer assumes this is the secret to survival in the corporate world, the results are often disastrous. Eventually one will face the consequences of one's actions, especially when the actions are repeated.

This prompts the question. How can bad behavior be necessary and when can it be used to one's advantage?

For this, Machiavelli said:

> I believe that this follows from severities being badly or properly used. Those may be called properly used, if of evil it is possible to speak well, that are applied at one blow and are necessary to one's security, and that are not persisted in afterwards unless they can be turned to the advantage of the subjects. The badly employed are those which, notwithstanding they may be few in the commencement, multiply with time rather than decrease.

According to Machiavelli, when committing a violation, the ruler must ensure it is necessary and not just a malicious act as per one's whim and fancy. Unfortunately, those in power are often corrupted by power to the extent that they believe they have the license to harass and intimidate their subordinates.

Additionally, companies that adopt a hire and fire policy often consider it an entitlement to use as they please rather than only as required for business needs.

Conversely, employees who indulged in bad behavior out of necessity are deluded into thinking it can be used as a strategy rather than as a last resort. This is what Machiavelli considers a poor use of wickedness. Instead of using an end-justifies-the-means philosophy, one loses sight of goal and considers the means an entitlement.

In the same manner, when an engineer commits a violation against his or her company, it needs judicious thought about whether it is necessary. In the case of ghosting described above, one could say such an action is justifiable if they had good reason to believe their current company would disrupt their future plans.

One usually fears that committing a violation may cause one to not just face legal action, but also result in a loss of credibility among their peers and colleagues which could affect their careers at a later stage.

Hence, if an engineer made an honest effort to leave the company in a professional and smooth manner and faced clear and repeated obstacles by the management, their actions would be justified. Furthermore, per Machiavelli's philosophy, a violation which ceases once is no longer needed, is a good strategy.

> *He who does otherwise, either from timidity or evil advice, is always compelled to keep the knife in his hand; neither can he rely on his subjects, nor can they attach themselves to him, owing to their continued and repeated wrongs.*

Machiavelli cautions against violations that do not cease, but rather increase as time passes.

In such cases, the parties injured by the ruler grow to a size when they can rise against the ruler who then must continuously resort to cruelty to suppress the people. A manager who fires employees indiscriminately and without reasonable justification will lead employees to recognize them as an egomaniac which will sooner or later result in an mass exodus from the company.

On the other hand, an employee who repeatedly jumps jobs while indulging in bad behavior or ghosting when no reason for doing so can be found, will eventually be branded as unprofessional, by both peers and managers.

In the previous chapters, I described how management will adopt various strategies of harassment as a tool to systematically suppress employees. In this chapter, one must differentiate acts of wickedness from acts of harassment. In the case of harassment, it is difficult to draw a clear line and label an act as unethical or illegal since any manager is entitled to use their perceived ethical judgement.

On the other hand, an act of wickedness is one that is either unprofessional, unethical or illegal. Machiavelli wrote many chapters on how a ruler is often required to be cruel and wicked, though he has also wrote extensively about how rulers must know the limits to which they can be cruel.

These discussions will at times overlap and might be difficult to differentiate, and therefore, in this chapter, these discussions are combined. Managers feel they are entitled to use their powers and the final decision rests with them. Nevertheless, every action taken by management is observed by employees, and unless it was justified, can result in the remaining employees making exit plans if they fear they will be next.

When speaking of violations either committed by the management or by an employee, the prospect of legal action always exists. Companies have the right to hire and fire employees at will, though, in many cases, the law requires the employer to fire employees only with justifications. In such a case, the company incurs a risk by assuming the employee will not take legal action. Except for the most egregious of cases, an employee will rarely seek to legally challenge their termination as the resultant publicity will harm their future career.

In many such cases, company management will violate the law as they assume legal recourse is too expensive and risky for an employee. One hears of high-profile cases where terminated employees take their companies to court and demand settlements. Though, in most cases, these are cases of members of the upper management being terminated. Such individuals possess the means to fight a long legal battle.

In exactly the same manner, companies can pursue legal action against an employee who willingly commits a violation. In the case of ghosting, or for that matter, violations to other clauses in the employment contract, a company can pursue an employee for legal damages. There have been a few high-profile cases where companies pursued ex-employees for various reasons such as violations of non-compete clauses, theft of intellectual property or for poaching employees.

Relatedly, in the reverse scenario, unless the violation by the employee was extremely damaging to the company, companies rarely pursue ex-employees. Moreover, if it appears that the ex-employee has no intention of continuing to damage the business of the company, a company risks negative publicity from taking legal action.

After dealing with how a ruler can use wickedness to their advantage, Machiavelli dedicated another chapter to whether a ruler should be loved or feared.

The simple answer, Machiavelli said, would be that a ruler should be both loved as well as feared, but when one must be chosen over the other, a ruler must be feared rather than loved.

> Because this is to be asserted in general of men, that they are ungrateful, fickle, false, cowardly, covetous, and as long as you succeed they are yours entirely; they will offer you their blood, property, life, and children, as is said above, when the need is far distant; but when it approaches they turn against you. And that prince who, relying entirely on their promises, has neglected other precautions, is ruined; because friendships that are obtained by payments, and not by greatness or nobility of mind, may indeed be earned, but they are not secured, and in time of need cannot be relied upon; and men have less scruple in offending one who is beloved than one who is feared, for love is preserved by the link of obligation which, owing to the baseness of men, is broken at every opportunity for their advantage; but fear preserves you by a dread of punishment which never fails.

The fickle and varying nature of rulers was prominent during Machiavelli's time when rulers constantly changed their alliances.

Beside rulers, there were also nobles who formed an ever-changing layer in the power structure present in those medieval times. As far as Machiavelli was concerned, a ruler must instill fear in the people and the nobles, such that if they did disobey or offend him, punishment follows.

When it came to describing the qualities of Duke Valentino, Machiavelli said *help him with zeal and offend with caution*. Promises were shallow if they were uttered by those who had nothing to lose by breaking them.

Therefore, Machiavelli believed a ruler could not rely on alliances that are made on the basis of promises and must instill fear in those who offer their support.

A majority of the readers who worked in any office can attest to the fact that most workplaces are nasty and toxic. Unless management is governed by visionary leaders creating an atmosphere of openness and collaboration, most companies are rife with factions of many forms.

As described in the previous chapter, such factionalism in a company can be fostered by the management as a strategy to put employees under constant pressure, not just to perform, but to also fend off one another. Nastiness can make its way from the very top of the company to the very lowest levels of employees who do not even have full-time contracts or benefits, and sadly, are the worst to suffer.

Nonetheless the inherent dark side of human nature plays a significant role in the toxicity of workplaces. The practical reason for allowing workplaces to become toxic lies in the fact that it allows companies to tip the conditions of employment in their favor and to the detriment of employees for the simple reason that it is much cheaper to coerce employees into working longer and for lower benefits with the ever-present threat of punishment than it is to encourage them to work, which might require incentives.

In the prior chapters, we examined how nepotism and harassment were used as strategies by companies and managers. When I speak of cruelty, this should be distinguished from nepotism and harassment. Cruelty refers to a fear of punishment instilled by the management in employees by various means, to extract a greater quantity of work from employees, despite being a clear violation of the existing employment contract or labor laws.

Unfortunately, in present times, there are many who celebrate overworking and take pride in the fact that they are under constant pressure, thereby making it difficult to label it as a nefarious practice. Hence, to be able to truly define cruelty in the corporate setting, one needs to adapt Machiavelli's philosophy rather than apply it directly, as the cruelty that was widespread in medieval times is nowadays found only in the rarest and extreme cases.

The working conditions of any employee are determined by labor laws and the employment contract between the company and the employee. Labor laws establish the bare minimum in terms of the well-being of workers and employment contracts will either stick to these basic requirements or provide additional benefits, if such is the norm in the industry.

The flexibility in contracts is usually determined by a mismatch between the demand and supply of workers in the domain. The greater the supply of workers, lesser are their benefits and less flexible are their contracts. On the other hand, in domains where skilled workers are scarce, companies continuously update their benefits and reword their contracts to stay ahead of their competitors.

Competition and basic human greed lead companies to find loopholes in labor laws and bend contracts in their favor, while chipping away at employee benefits.

This is the sad reality in almost every industry. A worker ends up working longer hours than they did five years back and though benefits may increase, they rarely keep up with inflation and the rising cost of living. There are many ways companies force employees to accept this gradual erosion, some overt and explicitly hostile, while others are subtle and barely noticeable.

These acts can be considered acts of cruelty similar to those described by Machiavelli where a ruler injures a citizen posing a threat to his rule by either taking the person's life or freedom.

A company's policies can be cruel because failure to accept them can result in either outright termination or an obstruction to growth and promotion.

The reason for this practice is obvious: to increase profitability. By extracting greater output from employees for the same or eroding benefits, the company's profitability will increase. When employees do not have ways to voice their opposition, it gives further license to the practice.

In the relentless chase for profits, companies use a variety of strategies to continuously move the goal posts for employees. The ever-intensifying rat-race is accepted as the reality of modern life. We compare it to medieval times when the common people were the serfs of the nobility and feel thankful that our conditions are at least better than they were before.

Some of us expected labor laws to be amended to stifle the continuous extension of working hours. As already stated, greed causes companies to find ways around labor laws. To arrive at a feasible and sustainable solution that works for employees, one should acknowledge violations can be committed by employees against companies as well. Similar to the salami-slicing strategy used by companies, employees can find ways to reduce their efforts to their companies as already described in the previous chapter.

Such an act by an employee can also be an act of cruelty since it is a willful violation of the employment contract. Most employees are unwilling to indulge in such a manner, as most of us are raised with sermons admonishing bad behavior of any form. Notably, in these modern times where profits are the be-all and end-all of human existence, one must find ways to reconcile with one's conscience.

The past chapters described some of the various ways an engineer can remain engrossed in challenging work despite being in companies where the work is mundane. An engineer seeking an existence independent of their company is a professionally and financially wise survival technique.

In this chapter, however, the discussion is related to how an employee can deal with management constantly finding ways to squeeze employees to the maximum. Nowadays, employees find themselves in this position and are either exhausted or completely burned out from meeting the demands of the management or attempting to negotiate temporary reprieves.

For many employees, this might be the only solution. Remember, it takes two to tango. Exploitation needs both exploiter and the exploited and the former disregards laws and ethics while the latter is unable to bend the laws or act unethically.

> *Therefore a prince, so long as he keeps his subjects united and loyal, ought not to mind the reproach of cruelty; because with a few examples he will be more merciful than those who, through too much mercy, allow disorders to arise, from which follow murders or robberies; for these are wont to injure the whole people, whilst those executions which originate with a prince offend the individual only.*

The above quote by Machiavelli describes how it's better for a ruler to be cruel and govern effectively rather than be kind and allow disturbances to arise. In a similar manner, for engineers, it is better to commit calculated violations if it helps their careers, rather than attempt to be a model employee at the cost of being exploited and manipulated.

One may think a cruel ruler is an unfortunate necessity for dealing with those who challenge law and order. In a similar manner, employees accept management exploitation—assuming the eventual profitability of the company will guarantee their positions in the company.

Unfortunately, this assumption is wishful thinking. Companies pursuing nefarious policies for profitability will have little thought for the welfare of the employees. On the other hand, an employee acting in a manner to hurt the company is perceived as selfish and petty. Nevertheless, there are direct and indirect benefits to employees when they place their interests before the interests of the company.

Before we begin to talk about benefits, it is important to describe what it implies for an employee to commit a violation against their company.

As earlier established, a company may not resort to disciplinary action or legal recourse for minor violations, though could do so for major violations. Therefore, one must acknowledge red lines one must not cross. To determine these, of course, is specific to the industry and the company, and also the company's competitors.

In general, a few examples could be provided as examples of major violations. Selling confidential information to competitors, working or consulting for competitors, engaging in parallel business activities that could damage the business of the company, excessive usage of facilities and resources of the company for work not related to the company and a few others, would be considered violations serious enough to result in termination, and even subsequent legal action.

Though a company would expect an employee to be solely working for the company and use the facilities and resources allotted to the employee solely for the work of the company, such rigid compliance is rarely enforced.

Let me now explain benign activities employees can use against management. In the previous chapters it was shown that every employee needs to invest a non-negligible proportion of time in activities that improve their professional profile, whether related to training and gaining advanced knowledge or on side-projects enabling one to hone one's skills while potentially opening the door to long-term revenue generating activity.

When an employee indulges in these activities at the expense of the company's work, the employee commits a violation that could be termed as a form of cruelty against management. Often, an employee indulging in external activities during the official working time of the company is labelled as dishonest, deceitful or petty. Such an employee is made to feel they are no better than a corrupt government official who squanders the taxpayer's money for unethical personal profit.

However, as explained, most employees are already at the receiving end of numerous unethical practices of a company and labelling them as the perpetrator for minor violations is hardly a fair comparison.

Employees need to remember that no one is irreplaceable and in this ruthless world driven by profit, companies will find ingenious ways to render their employees redundant or find ways to cut their remuneration and benefits.

Through a process of continuous training and side-activities, engineers can stay a step ahead of the rat race, if not avoid it altogether. In this chapter, the discussion will revolve on why an employee should indulge in these activities at the expense of the company, rather than exclusively on their own time as company management prefer.

Many companies have no objection to their employees indulging in training or side-projects unrelated to the business of the company as long as these are done in the employee's personal time and with the employee's own resources.

Though this might seem only fair, most companies found ways to gradually encroach on the personal time of employees. The international nature of modern work requires many employees to be available at times well outside the regular working hours of the company.

Some companies go so far as to state their philosophy that employees should not be *clock-watchers,* a term for those who disconnect outside of regular office hours.

Once again, due to the nature of modern work, companies expect their employees to remain available during their personal hours, may this be evenings or weekends. Sadly, as is obvious, this encroachment tilts employment terms in the favor of companies.

While some employees were persuaded to accept extended working hours and being asked to be available outside regular office hours as the new norm, most companies still cling to the sanctity of the office and the regular hours an employee spends in the office. Other than the fact that employees are allowed to take breaks, most companies strongly discourage their employees from using regular hours for any activity that is not related to the work of the company.

This is done through overt and covert means, some annoying while others are downright offensive and mean. Companies monitor the time spent by employees in the office when they swipe in or out at access-controlled entrances and exits.

They adopted time management software requiring employees to log the time spent on different activities. Some companies even use spy cameras and spyware installed on work computers to monitor the performance of employees.

Employees find themselves chained to offices where presentism scores over productivity with the expectation that one must arrive well before opening time and leave after closing time. It's silly to be treated like children sent to kindergarten by parents who need to work. The underlying philosophy of rigid office hours is similar, namely, to ensure that employees are chained to the work of the company for a minimum duration.

One could cite numerous arguments about a drastic increase in productivity when a flexible working hour policy is adopted.

However, many companies prefer the simplicity of rigid working hours as opposed to flexible working hours, as it can be an additional burden for the management to regulate. Following the arguments above, it is obvious the only solution is for employees to indulge in external activities within the confines of the office during regular office hours.

If an employee is engaged in training or an activity completely unrelated to the work of the company, to indulge in it during office hours within the confines of the office entails a certain level of risk. Even if the activity will not result in disciplinary action or outright termination, if one is observed by colleagues or supervisors to indulge in external activities, it creates a bad impression or a sense of distrust.

In most companies, the growth and prospects of employees are largely determined by the relationship the employee has with management and colleagues. If an employee behaves carelessly—such that it causes their colleagues as well as the management to lose their trust—their future in the company can be jeopardized.

Hence, indulging in external activities within the confines of the workplace should be done in an inconspicuous manner. Depending on the nature of the company and the facilities at their disposal, it can be easy or almost impossible to indulge in external activities without attracting attention.

In large companies with separate IT departments, an employee might find far greater restrictions, as resources provided to employees such as computers are installed with spyware that record the usage of the machine to a disturbing level of detail. Furthermore, access to the internet is restricted, personal email cannot be accessed and many websites might be blocked.

In such cases, employees need to segregate tasks of their external activities into those that can be performed in the office and those that can only be performed outside the office.

In smaller companies where employees are allowed a greater degree of freedom, one can indulge in external activities more freely as long as one does not utilize the company's resources to an extent that attracts attention, such as frequently downloading files or printing documents.

Most employees will hesitate to do anything other than the work of the company while within the confines of the workplace. This is partly due to our general upbringing that dissuades us from being dishonest or insincere, and partly due to fear of disciplinary action. To reconcile with one's conscience, one needs to remember that such dishonest behavior is a pushback against a gradual erosion of rights and privileges that companies got without significant opposition from workers.

If one allows their company to extend working hours while creating a competitive atmosphere that puts employees in a rat race for survival in the company, the career of the employee will be both agonizing and precarious. As Machiavelli suggested, a ruler cannot always be righteous and just, but must also resort to force or fraud when necessary; in these modern times, one cannot assume that honesty and sincerity will be rewarded.

The above is an obvious justification for an employee to be dishonest. However, there is a far subtler consequence of an employee's dishonesty which is difficult to comprehend unless one practices it. When employees are taxed and exploited to the point of being burned out, they lose interest in their jobs and usually become vocal in expressing a negative view of work in general.

As an example, if engineers in software companies are routinely overworked and are in toxic environments that put each one at another's throat, such engineers become outspoken critics of the software industry. The underlying fact is that all industries are similar and the condition of a software engineer is not different to a marketing professional.

Herein, by being vocal in their criticism of their domains, they paint a gloomy picture of the industry that would cause an outsider to think twice about entering the industry. If one questions any engineer in any industry about how they ended up specializing in that domain, often the answer will be that they found this calling either because of their family, friends or academic circles, or at times even a mentor who encouraged them to enter the domain.

For any profession to thrive, it needs a continuous supply of young talent.

This supply of young talent can only be ensured if there are professionals in the domain who are satisfied with their careers and see a promising future. The fact that a profession provides good remuneration or benefits alone is insufficient to ensure talent can be acquired and retained.

Unfortunately, most companies have no strategies in place to solve this problem—though most will acknowledge the severity of the problem. Over the past decade, many engineering domains face an acute shortage of talented engineers. The engineers in the domain are trying to exit or continue with their jobs solely due to a lack of other options and were not providing young engineers with motivation to enter the domain.

In many occasions, industry kicked the ball to academia, expecting colleges and universities to magically instill a love so strong for engineering in students, that they would willingly subject themselves to all forms of exploitation.

With myopic company management unable to see beyond profits and engineering employees unable to find a way past management, the stagnation of engineering has vast repercussions—not just for the economy but humanity in general. A scarcity of skilled professionals slows down innovation. One expects laws to be amended to acknowledge this problem and incentives provided to companies to retain talented professionals.

But as voiced before, human greed finds a way around laws and if companies find it easier to exploit their employees to maximize their profits, laws can't keep pace with human ingenuity.

To find a solution to this gloomy scenario, Machiavelli's philosophy can be adapted. In the same way a ruler can govern their subjects well and keep them content need not worry about being cruel under certain circumstances, thus skilled professionals should not allow their passion to fade in the face of corporate greed.

A skilled professional flourishing in a company while surreptitiously indulging in activities independent to the company will not only be satisfied with their career but will also serve as a role model to young professionals to enter the domain.

As students we are taught to sacrifice our personal wellbeing for the greater good. We carry this foolish lesson into our professional lives where we end up sacrificing our personal wellbeing not for the greater good, but rather for the profit of companies. The sacrifice we make is not for a noble cause, but rather for the greed of a select few who form the management of companies.

As a result, those whose services could have benefitted humanity are frustrated to the point of surrender and those who have benefitted would rarely put their gains to good use.

Nonetheless, as an engineer, it is prudent to indulge in independent activities while working for a company. By letting oneself get bogged down by the continuously increasing demands of management and losing our passion for our work, we not only let our careers stagnate, it causes stagnation in our professions.

If by deceit and dishonesty one can attain professional satisfaction and remain competitive in the labor market while fulfilling our obligations to our company to the extent that we retain our positions, not only do we flourish professionally, we are also role models for the younger generation of professionals.

> And above all things, a prince ought to live amongst his people in such a way that no unexpected circumstances, whether of good or evil, shall make him change; because if the necessity for this comes in troubled times, you are too late for harsh measures; and mild ones will not help you, for they will be considered as forced from you, and no one will be under any obligation to you for them.

This obscure statement by Machiavelli is difficult to comprehend at first. Herewith, all it implies is that a ruler should not constantly change with times, but rather have a consistent policy of governance that shows wisdom and courage. The cruelty necessary to put down those that threaten the state should not be exercised according to whim and fancy or by the need to appease or terrorize, but rather well thought out actions that preserve the state.

A ruler who resorts to cruelty and violence only in hard times will be perceived as a weak ruler who cannot summon the courage to do so otherwise, and therefore, those harsh measures may be of little use. On the contrary, a ruler who seeks to appease in face of a rebellion might be seen as a weak ruler and will face greater rebellions in the future.

In a similar manner, an engineer ought to study management actions closely and differentiate between a management that implements policies that ensure profits and a management that is driven purely by greed. A management that resorts to harsh actions only during times of distress is a management that cannot be guaranteed to keep a company afloat.

In contrast, a management adopting a balanced approach of offering incentives and meting out punishment with a view of ensuring the smooth operation of their company, is a management an engineer can place faith in. Likewise, an engineer should strive to be consistent in their behavior while working steadily and meticulously and not needing the constant fear of punishment to deliver on projects.

An engineer should invest sufficient time on their own improvement independent of the company and not wait for the moment when they might face termination, as by that time, if the foundation has not been laid, it will be too late to prepare for a smooth exit.

> *Nevertheless he ought to be slow to believe and to act, nor should he himself show fear, but proceed in a temperate manner with prudence and humanity, so that too much confidence may not make him incautious and too much distrust renders him intolerable.*

Machiavelli offers a word of caution to a ruler.

A ruler must be seen as wise and prudent and not rash and indecisive to instill confidence in the nobles and the subjects. A ruler who commits violence without deliberation will hurt enough people to rise against him.

The same can be said about the management of a company.

Managers should be meticulous and decisive and not rash and bullying. A manager who survives by dividing their teams and harassing employees will eventually not find sincere workers to deliver during times of distress. Unfortunately, anyone who is in a position of power assumes they have the right to harass and intimidate subordinates. Often such unbearable managers lead to an exodus of employees. Prudently using power will ensure equitable treatment of employees to achieve a profit, while using punishment only when it is justified.

The same can be said of engineers.

An engineer must not become reckless and casual about work, nor be blindly devoted to the company. An engineer must behave in a way so management remains confident and assured about their abilities. An engineer must not place blind faith in the management, or even hope management will protect them during hard times.

An engineer should extract a portion of time commensurate to the increasing demands of management to use for their independent growth, avoid being exploited by management and forge an independent existence.

This chapter would be one of the first in the book that combines several chapters of Machiavelli's *The Prince* dealing with similar content, namely how rulers should act to preserve their kingdom in ways that may not be deemed to be commendable.

The chapter includes one of Machiavelli's most (in)famous quotes: ...*it is better to be feared than to be loved.*

Many of Machiavelli's chapters were considered shocking at the time of its publication, and some of the contents of this chapter might be considered to be unethical and condemnable.

That said, anyone who has spent a significant amount time in industry or as a graduate student in universities will find the contents resonating with their own experiences.

This chapter used Machiavelli's analysis to describe the fickle and ruthless nature of the modern corporate world. The same companies that proudly welcome employees into their *families* will turn around and fire them without a second thought.

The same companies who scream slogans of work-life balance formulate policies to push their employees to the point of burn-out.

In the same manner Machiavelli suggested a ruler become a totalitarian and rule by decree, companies strive to maximize profits with little thought for anything else. Human greed is in a different form in these modern times. To satisfy their greed, corporations bend, break and subvert every law and act in unethical and contemptible ways.

Strategies used in corporations can be equated to Machiavelli's philosophy about how rulers should consolidate. For most of us engineers who deal with various levels of managements, one can refer to how Machiavelli believed rulers should behave when they have been placed on the throne.

One can find parallels in how Machiavelli believed Cesare Borgia performed in an exemplary manner with how many managers behave when they are appointed to manage a team.

From this analysis one can find ways that an engineer can survive when led by managers who in turn are only content with surviving at any cost.

When one is dismayed at how corporations and their managements behave, it becomes clear how wickedness is used to consolidate power.

At times like these, we are frustrated at how employees have few options for dealing with a management that behaves like a bloodthirsty feudal lord.

In consequence, this is a power struggle attempting to break through obstacles placed by legislation. Companies behave in shocking ways to maintain the status quo in their favor, assuming that employees will complacently toe the line. As brought up by a few examples, employees behave in a wicked manner to break the stranglehold companies have.

Machiavelli believed that—except for the hereditary ruler—few could escape the reproach of cruelty. He believed it was better for a ruler to be cruel than for a ruler to be weak and allow the kingdom to plunge into chaos. He provided some examples to illustrate this point.

Most of us who have worked in companies have sadly accepted the fact that no business can survive in this modern world while being ethical and humane. In the relentless drive to maximize profits, companies squeeze their employees to the maximum while offering only a minimal increase in benefits.

Employees accept this continuous erosion of personal space in exchange for a job. Nonetheless, in the same manner companies pursue a policy for maximizing their profits, employees can also find ways to maximize their employability at the expense of their companies.

A Climb to Management

THIS CHAPTER DRAWS from a single chapter titled *Concerning a Civil Principality*. In this chapter, Machiavelli talks about how at times, in the case of a territory that achieved independence, the nobles and the citizens were at loggerheads on who will be the next ruler.

A potential outcome of such a power struggle is a civil principality where the ruler is chosen from among the common people. This chapter applies to the modern world, as engineers will find themselves promoted to management at some point of time.

The path to becoming a good manager is not the focus of this book, but Machiavelli's writings will be examined to guide an engineer who might embark on a transition to management.

> I say then that such a principality is obtained either by the favour of the people or by the favour of the nobles. Because in all cities these two distinct parties are found, and from this it arises that the people do not wish to be ruled nor oppressed by the nobles, and the nobles wish to rule and oppress the people; and from these two opposite desires there arises in cities one of three results,

either a principality, self-government, or anarchy.

If among the nobles, no one emerged powerful enough to become the ruler and the rivalry between them made it impossible for one of them to hold on to the throne, this provided an opportunity for new possibilities. Of course, such a territory could've become a republic, which is what Machiavelli implied by self-governance, or could descend into chaos in the absence of good leadership.

Per contra, if it were to remain a principality, a leading citizen could break the deadlock between the nobles and the people by ascending the throne.

The challenges faced by a new ruler are massive. Not only are there many of the usual challenges of statecraft, but the new ruler is a novice amid nobles already well acquainted with the game.

In the corporate world, it is natural for employees to aspire to a managerial role, as purely technical roles offer limited opportunities after a certain stage. Only a few mega corporations can offer purely technical growth paths for employees who want to remain purely technical throughout their working lives.

In other cases, in an incomplete transition to management, one is offered roles combining management and technical expertise. Thus, transitioning a technical person into a management role does not require a conspiracy between the management and the employees, but is a natural process.

Having been in industry, I have seen several cases of engineers suddenly acquiring management positions. A few did well, while most were disasters. I admit I have very limited experience in this domain, as I have spent only a short period in a management position with mixed results. Be that as it may, combining my own experiences with my observations of managers during my years in industry, I can write a bit on the transition to management that many engineers contemplate at some point of time during their careers.

Discussions in the past chapters are relevant when

considering how a manager should behave if examined from the perspective of being a manager trying to survive in a company.

When one transitions from an engineering role to a managerial role, it is important to note that one has entered the company's power structure, even if it is only at a low level. In the past chapters, the discussion was about how a powerless, subservient engineer can survive in a company.

In this chapter, the discussion will change as the new manager now has power over those under their supervision. Moreover, the discussion is now more complex, as the manager is in turn subservient to those in upper management, besides having rivals who are other managers at the same level.

Appling Machiavelli's philosophy to this situation, I break up the discussion for understanding into three stages:

- *Dealing with subordinates*
- *Dealing with peers*
- *Dealing with superiors*

> *It is to be added also that a prince can never secure himself against a hostile people, because of their being too many, whilst from the nobles he can secure himself, as they are few in number. The worst that a prince may expect from a hostile people is to be abandoned by them; but from hostile nobles he has not only to fear abandonment, but also that they will rise against him; for they, being in these affairs more farseeing and astute, always come forward in time to save themselves, and to obtain favours from him whom they expect to prevail.*

Machiavelli says that when a private citizen is made ruler with the support of the people rather than from the favor of the nobles, the new ruler need only ensure that the people are not oppressed during his rule in order to continue to enjoy their support.

On the other hand, when a private citizen is made ruler by the nobles, he must win over the people by protecting them. In this chapter, Machiavelli's writings are milder as compared to other chapters where he openly vouched for tyrannical violence. Consequently, the common thread in these chapters is how a ruler secures themselves internally before worrying about foreign invaders posing an external threat. It would be very difficult for a ruler to defend their kingdom when the people are hostile.

In a corporation, a manager is generally appointed by the upper management and can be removed if they lose their confidence. Only in rare cases do the employees have the power to remove or prevent the removal of a manager. Other managers, on the flipside, can create a hostile environment for the new manager and make it difficult for them to survive.

There are many reasons why a manager might lose the confidence of the upper management and be removed by them.

Understanding these reasons is far beyond my analytical capability. Moreover, a detailed description of management is beyond the scope of this chapter which is restricted to the specific case of a talented engineer rising to a management position where the new manager possesses a set of skills that can still be used in this new non-technical position.

For a new manager, to retain one's position, interaction with subordinates must lead to an acceptable quality of work produced by the team. A manager is responsible for their team achieving the goals and objectives set for the team by the upper management.

Social media is filled with posts of managers with superhuman capabilities. Unfortunately, such posts set false expectations for a manager, which in many circumstances cause them to burn out. Except for the few cases where the manager ends up being a co-owner of a company, in a vast majority of cases, the manager is just another employee and the final objective should be to secure one's position while establishing a record from which higher positions can be attained. Therefore, a new manager must be wary of attempting to set incredibly high standards of leadership, but rather transition and grow gradually in the role.

The human factor plays a major role during the beginning of a new manager's tenure, as the inherent leadership qualities of an individual come into play at this point. Some are more inclined to be leaders and command respect, while others cannot control their team. A management position can be a completely new one, where the upper management decides to bring in someone completely new to the organization, or can be an internal appointment, where someone within the organization is elevated to this position.

As a new manager, one should apply one's newfound authority to the benefit of oneself and one's team. This is much easier when the new manager is with a team they haven't worked with before. Contrarily, if one supervises a team one was a part of previously, it is much more difficult to deal with them as subordinates and for them to deal with the new manager, their ex-teammate, as a superior.

A new manager often mimics managers who were their peers, which unfortunately yields poor results. In a company rife with favoritism, a new manager might accept that as a norm and apply that as well. As already stated, beyond fulfilling the shallow and petty nature of managers, nepotism plays a deeper role in suppressing talent while establishing mediocrity as the baseline of the team.

When an engineer transitions to a managerial role, he or she still has an engineering background, and when setting the baseline of the team, this baseline should be with respect to one's own skills and talent. Setting too low a baseline creates an under-performing team, while setting too high a baseline can jeopardize one's own position.

A new manager needs to spend time assessing the capabilities and nature of the team members who can be divided into several categories. The first are those who are skilled and talented and possess the drive to achieve. Often, despite skills and talent, they possess mediocre or poor communication and social skills.

The second are those with mediocre skills and talent, but with good communication and social skills. The few other groups are the minority groups—skilled and talented who have equivalent communication and social skills, and the last group of the ones with poor technical skills and equally poor social skills.

The first two groups are those a manager can use to advantage, while the last two groups are those that a manager must be wary of.

Before we examine how favoritism or nepotism can be used to one's advantage, one must acknowledge that a utopian system of equal opportunities is sheer fantasy. Resources can never be allotted equally to everyone, and even if it can be achieved by any stretch of imagination, such a system will be inefficient and wasteful. Those who possess greater abilities will not have adequate resources, while those who are incompetent will waste them.

Therefore, as a manager, one needs to allot resources according to the ability of the members of the team and not randomly. Inadvertently, this leads to some form of nepotism. *Will the nepotism benefit or hurt the team and the manager in the long run?* It is then a question of how it is applied. In one case, it could result in a high performing team with a good deal of contentment among most members of the team, while in other cases, it could result in a mediocre team with frustrated and disinterested engineers.

In the case of a skilled engineer transitioning to management, it would be obvious to allot greater resources to the talented engineers and allow them to work with a certain degree of freedom. This is particularly the case when the manager has a technical background in the recent past which team members acknowledge.

As one spends time as a manager, technical skills gradually erode to be replaced by organizational skills. In the beginning, the manager has the advantage of being a technical person who commands the respect of the talented members of the team, who might otherwise have looked down on a purely organizational leader.

One of the greatest risks to giving a free hand to talented engineers comes with the fact that most of them with poor social skills would otherwise use this freedom to become disruptive. To begin one's stint as a manager with a high-performing team provides a great deal of momentum and creates a favorable first impression on upper management.

A new manager with a strong technical background might be inclined on the other hand to immerse oneself completely in the work of the team, and work alongside the team. This continuation of one's previous background will result in a neglect of the organizational duties expected from a manager and create a disconnect with upper management. In this situation, a manager can utilize the second group of team members, namely those with mediocre technical skills but good social and communication skills.

By delegating organizational duties and assigning project monitoring roles, one can utilize their skills while keeping them content. To enhance the morale of the team, the manager should ensure talented engineers are not commanded by the mediocre but enjoy a degree of independence—as long as they remain loyal and support the manager.

This might seem like those idealistic social media posts, but there are challenges to maintaining such a team. The biggest lies in the fact that in most companies, a large proportion of the work is mundane and though it is essential for the profit of the company, it's not challenging to a talented engineer.

Most managers focus only on the goals and objectives of the team and will not cater to talented engineers and how they should be kept occupied. If left unchecked, the mundane nature of work will cause talented engineers to be frustrated and discontented.

To remedy this, revisit the solution presented in the previous chapters, namely for talented engineers to pursue interesting side-projects.

In the past chapters, the discussion was written from the perspective of an engineer surviving in a company. Since this chapter is written from the perspective of a new manager, the question to be asked is—*how can a new manager encourage an engineer to pursue a side-project?* In the few companies with this open approach, management put in place a progressive policy about side-projects. Some companies explicitly state in their employment contracts that any work done by an employee outside the domain of the business of the company remains the property of the employee. The only condition is that the employee notify the management in the event of a potential conflict of interest, where the side-project could infringe on a product or service offered by the company. Such policies are welcome, and whether enforced by law or voluntarily imposed by the company, encourage innovation and creativity from employees.

Despite policies adopted by a company, it is an engineer's immediate manager who decides how easy it is for an engineer to pursue a side-project. Many managers take a negative view of employees undertaking any other work for fear it will distract from the company's work.

As an engineer transitioning to management, one is in a strong position to take a different view, especially if one used to actively indulge in side-projects previously as an engineer. If a manager encourages and supports the engineers in undertaking side-projects, a vast number of benefits can arise that will help to consolidate one's leadership of the team.

As an engineer, one is aware of the impact on one's professional satisfaction and mental health when fully engrossed in a challenging and interesting project. As a manager, one should maintain an open channel of communication with the talented members in the team where they can discuss their ideas and you can offer suggestions.

It is important as a manager to draw the line between the business of the company and the ideas of the team members, and not allow the official work of the company to flounder.

For all that, even a ceremonial pat on the back for good effort and original thought can help a manager earn the respect of the team and keep them loyal. This loyalty will bear fruit in ensuring the deliverables of the team are achieved. Engineers will put in a good effort knowing that—under this leadership—they will enjoy freedom and independence.

Hence, for a new manager transitioning from an engineering background, the greatest challenge is to ensure that the team remains driven and focused while building loyalty in the team. The productivity of the team can be ensured by allowing the talented members of the team to work independently with the minimum amount of supervision necessary for project deliverables to be met.

As long as the talented have high morale, achieving deliverables should not be a challenge due to the fact that in most companies, disparity exists between the skills needed and the skills available. If one avoids an atmosphere where talent is belittled and sidelined, a high-performing team can be achieved.

As one gains experience as a manager, learn to craft strategies to deal with various conflicts as they arise. However, at the onset, if a manager fails to secure the loyalty of the talented, it is difficult to win it later.

> *Therefore a wise prince ought to adopt such a course that his citizens will always in every sort and kind of circumstance have need of the state and of him, and then he will always find them faithful.*

Machiavelli made the statement above in the context of magistrates and other officials within the principality who hold various offices. It's natural to delegate power to such officials or magistrates and it is important that the people see them as public servants, and owe allegiance to the ruler, and not to these officials.

Allowing localized pockets of power to develop can be dangerous, as in the face of an external threat, these magistrates can lead a rebellion against the ruler.

In the same manner, a new manager must ensure the team feels the need for the manager to continue. For different members of the team, such a need can arise from different reasons. For the genuine engineers in the team, it may be that the new manager allows them the freedom to pursue their side projects, while for a few others, they desire assignment to non-technical responsibilities.

In this manner, the new manager can prevent localized cliques forming within the team by ensuring that the team views the manager as their ultimate supervisor.

When a private citizen becomes a ruler through either the favor of the people or the nobles, Machiavelli advised remaining at guard against the nobles. When a ruler is formed through the support of the people, he has to contend with the open hostility of either all or a majority of the nobles, while even when one becomes a ruler with the support of the nobles, one must be wary of rival factions waiting for an opportune time to wrest power.

In a mid-size or a large corporation, with several managers at the same level, there are similar power struggles and even if other managers do not conspire directly to remove a new manager, they could still make it difficult for a new manager to function to the extent they could lose the confidence of the upper management.

Rivalry between managers usually boils down to each one attempting to project their team as the best performing team to obtain the best possible appraisals for themselves and secure future promotions. To do so, managers attempt to grab the best projects for their teams, or wrest projects away from other teams.

Such attempts can appear in various forms—refusing to share information, conducting parallel and competing work to propose very similar projects, or finding fault and discrediting another team's work. Besides such professional rivalry, there also arises human factors, especially when the new manager comes from a very different background with respect to age, ethnicity, orientation or gender.

The best antidote for professional rivalry is for the new manager to ensure that the team's performance remains high. This was discussed in detail above. Though there have been cases when a new manager possessed great social and political skill to sustain themselves with even a poor functioning team. For someone new to management, poor performance and lack of results can lead to failure.

Besides the quality of work, the new manager needs presentation and debating skills to favorably publicize the team's work. For many new managers, this is a challenge. Engineers deal with verifiable facts and data, but managers weave results into impressive stories.

A new manager can use training. For engineers with little experience with presentations and public speaking, it is advisable to begin with trainings on Effective Communication and Presentation Skills. As one gains managerial experience, the quality of presentations will gradually improve. It is important for a new manager to remember the difference between presenting as an engineer versus presenting as a manager, since the former usually presents data and results to facilitate the progress of projects and the latter uses presentations to convince and reassure the management.

One should avoid outright fabrication as it will damage one's credibility, but presenters should consider what the management wants to hear and not just what it should hear.

> ...nobles ought to be looked at mainly in two ways: that is to say, they either shape their course in such a way as binds them entirely to your fortune, or they do not.

Whether a ruler is made by the support of the citizens or by the support of the nobles, it would be rare cases for the new ruler to have the support of all the nobles or to have the support of none of them.

Those in power, over time, will have rivalries that will cause at least a few to favor a new ruler. In a similar manner, a new manager will rarely have the support of all other managers in the company and will also rarely find them all against them.

The new manager must divide other managers into two categories—potential supporters and potential rivals. The word potential is used, as affinities change with time, just as it did in Machiavelli's world.

> *Those who so bind themselves, and are not rapacious, ought to be honoured and loved...*

The statement above is obvious even without the specific case of a new manager in a company. In addition, there are many situations where those that appear to be supporters either may not be so, or may not be truly genuine. When a new manager has a background different from the other managers in the company, this evokes a multitude of reactions.

There will be those intrigued by the challenges and think management could be a refreshing change. In contrary, there are those who will be dismissive and feel these changes are experimental and will inevitably fail. The challenge arises when the new manager is among those who are experienced and are well-versed in putting forth a diplomatic standpoint which may not reflect their genuine positions.

> *...those who do not bind themselves may be dealt with in two ways; they may fail to do this through pusillanimity and a natural want of courage, in which case you ought to make use of them, especially of those who are of good counsel; and thus, whilst in prosperity you honour them, in adversity you do not have to fear them.*

The above statement by Machiavelli is difficult to interpret but could be viewed as a ruler *feeling the pulse*. Whether we consider nobles, renowned citizens in a principality, or managers or senior employees in a company, this group could refer to those who have accumulated a significant amount of experience, but through lack of ambition or simply avoiding responsibility who prefer not to be active.

Whether these people offer support or oppose, they vacillate and are unreliable. Counting them as supporters would be a grave error. Subsequently, instead of dismissing them—which one may be inclined to do if one views the for-or-against question as purely binary—these people can be used to gather historical information about the company and other managers.

This can help a new manager interpret the behavior of other managers and ascertain how genuine they might be in their support or how hardened they might be in their opposition, and the potential circumstances that might change their stance.

These wise people need kind words and a pretense of respect for their wisdom and experience.

> But when for their own ambitious ends, they shun binding themselves, it is a token that they are giving more thought to themselves than to you, and a prince out to guard against such, and to fear them as if they were open enemies, because in adversity they always help to ruin him.

This last group represents those whom a new manager should guard against, for they have experience and capability to be formidable opponents.

How one can deal with them is specific to the situation.

Here I suggest it is best that the reader seeks specialized content on conflict resolution. When faced with opposition from many formidable opponents, a new manager must rapidly acquire leadership skills to ensure their team performs well, and so they can retain their position in the company.

In my experience, few who entered such competitive spaces

were able to rise to the occasion. Since this is a worst-case situation, it is advisable that someone wishing to be a manager should prepare before accepting a managerial position rather than be overwhelmed after acquiring the position.

Machiavelli does not talk about how a ruler should deal with superiors, as in his analysis, a ruler is the supreme authority in a principality. However, in a corporation, a new manager will find an entire power structure to contend with, thereby making it essential to understand the situation.

For a manager, keeping upper management content to retain a position is more complicated than an engineer retaining his or her position. As an engineer, one is primarily responsible for the technical tasks assigned, while a manager is responsible for the performance of an entire team.

Surviving under a hostile upper management can result in mid-level managers experiencing burn-out as one feels pulled apart from both sides.

Some of the incredible levels of pressure felt by mid-level and entry-level managers is due to idealistic notions of management presented in social media. The constant need to communicate, organize, streamline and reassess along with other buzz words thrown around result in a manager's job never coming to an end. The manager cannot find a moment of peace.

One should read Machiavelli's philosophy, where despite delving into the complexities of statecraft, he breaks down leadership into an understandable and comprehensible set of guidelines. As one becomes a manager, one can strive to be a perfect manager and risk a nervous breakdown, or one can learn from Machiavelli's philosophy and choose a sustainable career path and allows one to enjoy one's leisure as well.

As already stated in previous chapters, a job is simply an exchange of skills and energy for compensation and benefits. This applies to a manager as well. A manager is responsible for ensuring their team performs to a degree acceptable to upper management.

As one ascends a company's hierarchy, responsibilities grow. Moreover, setting goals and objectives that constantly expand and change increase the burden without equivalent benefits. When dealing with upper management, the manager needs to be on guard that their responsibilities do not expand to include those of the upper management, as then one takes on another's job. This often happens in the same way a manager delegate managerial duties to engineers, as it is the natural tendency of those with power to extract more from subordinates.

In conclusion, becoming a manager requires a different set of skills when compared to being an engineer. As one grows in a company, the skills gradually become less technical and more organizational.

At this point, to conclude this chapter, I suggest that those who wish to pursue this path avail themselves of any number of books on this subject. My purpose of writing this chapter was to examine management from a non-idealistic angle, in the same way I examined the work of an engineer in previous chapters.

The idealistic narrative found in other mainstream outlets is either nearly impossible to achieve, or if even achievable, will require a superhuman effort.

Technical Weapons

THIS CHAPTER IS a combination of excerpts in several chapters in Machiavelli's writings—*That Which Concerns a Prince on the Subject of War*, *Concerning the Way in Which the Strength of all Principalities Ought to be Measured*, *How Many Kinds of Soldiery are There*, and *Concerning Mercenaries, and Concerning Auxiliaries, Mixed Soldiery, and One's Own*.

Though this part of Machiavelli's writings has not received much comment, nor has been quoted significantly, they are a vital part of Machiavelli's philosophy as this is one of the primary problems that infested Italy during those medieval times.

In my book, I changed the order of the reference to the chapters to make the material easier to digest for the reader, as I found myself going back and forth between the chapters of *The Prince* when I was trying to write this chapter.

The reader can, of course, read the chapters listed above in the order in which they appear in Machiavelli's book.

Machiavelli wrote extensively on warcraft and the problems plaguing Italy during those times. The past few chapters already cited some of the historical events that occurred, and which Machiavelli referred to in his book. At the time of writing his book, Italy was divided into factions and was without a single strong unifying command, something that Machiavelli sorely lamented.

However, a deeper analysis of his writings reveals the problems faced by Italy were far more deep-rooted than disunity. In his writings, he compared Italy with many of its neighbors, which he felt were better governed and better armed. Modern warfare rendered a lot of the discussions in Machiavelli's writings redundant if one read those chapters purely from a militaristic angle.

However, these chapters provide useful insight into how engineering companies can survive in this competitive world and also how engineers can retain their positions despite numerous factors at work to render them redundant. It is for this reason that this chapter is titled *Technical Weapons*. In our modern times, knowledge and skill have superseded brute force.

> *A prince ought to have no other aim or thought, nor select anything else for his study, then war and its rules and discipline; for this is the sole art that belongs to him who rules, and it is of such force that it not only upholds those who are born princes, but it often enables men to rise from a private station to that rank.*

This opening line of the series of chapters sets the tone for Machiavelli's core beliefs on warcraft. The above statement would in modern times be applicable only to military dictatorships, but in those times, legendary rulers were able warriors. Popular examples include Alexander the Great, Julius Cesar, Hannibal and a few others.

In the previous chapters, I quoted cases such as Agathocles the Sicilian and Oliveratto da Fermo who—despite being born outside nobility—rose through the ranks to become rulers due to their accomplishments in the army.

One might, however, feel it unnecessary for a ruler to be

militaristic when all that is required to remain a ruler is good governance.

For this, Machiavelli has the following argument:

> *Because there is nothing proportionate between the armed and the unarmed; and it is not reasonable that he who is armed should yield obedience willingly to him who is unarmed, or that the unarmed man should be secure among armed servants. Because, there being in the one disdain and in the other suspicion, it is not possible for them to work well together.*

During Machiavelli's times, the expectations from a ruler were very different from our expectations from modern heads of states.

During those times, when a ruler was not directly accountable to the people, though could face armed rebellions besides also having to fend off attacks from hostile states, Machiavelli believed that a ruler must be skilled in warfare.

A ruler who was not a warrior would not command the respect of his army, and Machiavelli gives countless examples of rulers who were murdered by their armies. During those medieval times, human rights were nearly non-existent and armies were infamous for horrific cruelty towards their own people and to those who have been conquered.

Thus, to command respect required the ruler to be a ruthless and skilled warrior.

In modern times, heads of states rarely have a military background, especially in states with some semblance of democracy. One expects the head of state to organize their government to function to an extent that they have the confidence of the parliament.

Besides organizational abilities, one expects a head of state to project an image as a leader the people have confidence in to lead the state, which is something that Machiavelli wrote about and which will be covered later.

In the modern corporate world, expectations of CEOs and

managers are similar to the expectations from modern politicians. To pinpoint a skill absolutely essential for any manager or company executive to possess, it's organizational skills.

Because this book is targeted towards engineers and not managers or leaders, it will not deal with the means by which a manager can acquire the organizational skills needed to lead an engineering team. There are numerous resources available for this purpose.

What *will* be dealt with is how an engineer should judge their manager, or for that matter, how the management of the company can be evaluated. As already expressed, poor management will eventually result in the collapse of a company, which affects all employees, with the worst affected being the most junior.

While employees can't do much about weak management, an engineer can devise effective survival strategies to ensure their career flourishes in a poorly managed company where even if it does not run to the ground, offers limited potential for growth.

An engineer must remember that a manager is responsible for the final deliverables of the team that they lead. It is the manager's responsibility to ensure that the team has all the resources necessary to perform tasks within a time frame. Towards this end, a manager must be proficient in project management tools with the objective of monitoring personnel, resources and time.

Though there are some naturally gifted organizers finding themselves like fish in the water in management roles, management in modern times is not just a human trait but a well-established and continuously expanding domain of study. Here, one can draw a parallel to Machiavelli's insistence on a ruler mastering the subject of war.

Like the fact that being a soldier was and is an established domain of study requiring one to continuously practice and innovate, being a manager is also a domain that needs continuous practice and replenishment, as the constantly changing world of business demands a new perspective every now and then.

In all my years in industry, I found myself working with various sorts of managers. There have been those who were primarily engineers who after being elevated to the post of managers, did not behave differently from engineers except for the fact that they were responsible for getting work done rather than themselves working.

In many cases, such managers were a disaster in the same way a regular civilian might think it easy to pick up a gun, take aim and shoot. A manager who believes their job is to talk and hold meetings is unlikely to be a good manager. Effective management is more than soft skills and the ability to conduct meetings. Bad managers let projects go off the rails, miss deadlines, are unable to extract work from employees and worse, blame their subordinates for what is clearly mismanagement.

A good manager need not be someone who speaks well but is meticulous and knowledgeable about the processes involved in the completion of a project or a product. Many engineers scorn managers who are not technically proficient, but this misunderstands the subject of management. A manager need not be a technical expert. Expecting technical expertise from a manager is unrealistic. A manager is not just responsible for the technical aspects of a project, but also financial and personnel aspects an engineer might not be familiar with. Every industry has specialized processes. Effective management requires a manager to be well versed with them, which in turn implies a continuous need to learn and refresh one's knowledge.

Before continuing with other arguments Machiavelli presents, let's change our perspective and examine the arguments presented from the angle of the engineer.

In our analogy, the engineer is the armed soldier led by a prince who acts as the commander-in-chief of the army. To command the respect of the soldier, Machiavelli urges the prince to learn the art of warcraft.

Let's now ask the question—*what must the soldier do to ensure that the prince protects, rewards and honors the soldier?* In the case of a soldier, all one can expect is for the soldier to be skilled, fearless and loyal and hope to be rewarded by the prince.

In an army, there is minimal room, if any, for independent thought and action, with discipline being rigorously enforced.

In corporate life, expectations of an engineer are different. An engineer has a certain degree of freedom to carve out an independent existence. This train of thought opens an interesting application of Machiavelli's philosophy for engineers in dealing with their managers.

As already mentioned, an unfortunate lesson taught to engineers is that once in industry, there is no need to augment technical skills. Instead, we should focus on soft skills. There is no denying that learning soft skills is useful and can help someone progress. However, the blanket statement that it is impossible or useless to augment one's technical skills is extremely dangerous and can lead to the disempowerment of engineers.

Like a soldier, an engineer's greatest assets are their technical skills and knowledge. Though a company may offer only a limited area for application of that skill or knowledge, they remain an engineer's prime assets. In comparison to a soldier, one could state that in an active conflict, not all of a soldier's skills are useful. But the greater the array of skills the soldier possesses, the greater will be their value to an army.

Therefore, the fact that the products or services of a company require only a narrow range of skills and knowledge is no excuse for an engineer to limit themselves. An engineer should never consider their present position to be their only or final position, as change might occur due to necessity and not through one's own choice.

Reference to the way Machiavelli urges a prince to make warcraft his primary domain of study, an engineer should make the acquisition and refinement of technical skills their primary objective. A skilled engineer will always find opportunities in any job market, and continuous learners will find ways into newer, emerging industries.

Those who complain that their skills are not utilized in their company might ask—*what use are these skills if they are not applied?* The answer has already been provided and lies in the previous chapters—*by undertaking side-projects.* Such side-projects are excellent sandboxes where one can hone and perfect skills and will eventually either serve as additions to one's professional portfolios or become independent ventures on their own.

There will be many obstacles to an engineer gaining technical skills. Beyond the constraints of time and energy, companies will use tactics to dissuade engineers from augmenting technical skills. Some of these tactics include burdening engineers with multitudes of useless, petty tasks which do not have much use or value but keep the engineer busy and exhausted.

Other tactics include companies promoting soft skills or non-technical events, thereby giving engineers the impression that pursuing technical excellence is unimportant.

Companies will also use tactics such as harassment and nepotism to frustrate engineers into losing interest in the technical aspects of the job. These were discussed in the previous chapters while examining Machiavelli's strategies.

A single answer to all these strategies is that if an engineer convinces themselves that technical skills are the answer to both a fruitful and sustained career, little can be done by any company to thwart their goals. Solutions exist for the nefarious strategies used by companies.

As already stated, in any company, there are tasks critical to the products and services offered by the company, while the remaining tasks are peripheral. To be a valued asset of a company, an engineer should ensure critical tasks are performed with diligence.

For the remaining peripheral tasks, one can use the strategy of salami-slicing and multiplex them with tasks in side-projects.

An engineer faces a low risk of disciplinary action if peripheral tasks are delayed or diluted. As for nepotism and harassment, solutions were discussed in the previous chapters.

Most important of all, and something that remains unclear, is that in the same manner that it is unreasonable to expect an armed servant to be obedient to an unarmed prince, it is also unreasonable to expect skilled engineers to devote themselves to their jobs when under the supervision of incompetent managers.

As an engineer, if one finds oneself getting frustrated at the ineptitude of the management, one must stop and ask oneself the question—*why am I even investing this time and effort?*

As understood and expressed, to waste one's precious effort and time in a poorly managed company will eventually hurt the engineer. When the company flounders, employees are left to fend for themselves.

> *As regards action, he ought above all things to keep his men well organized and drilled, to follow incessantly the chase, by which he accustoms his body to hardships, and learns something of the nature of localities, and gets to find out how the mountains rise, how the valleys open out, how the plains lie, and to understand the nature of rivers and marshes, and in all this to take the greatest care.*

Machiavelli urges a prince to constantly train, be ready for war and to practice leading his army into battle by learning the strategic aspects to war. In a military context, the above advice is natural and needs no explanation. Army units need to constantly drill and be prepared for war, as it is not just about being fit, but success is based on the execution of coordinated strategies during conflict.

Unfortunately, in the corporate setting, this advice cannot be directly translated, especially in the context of engineering teams. Sadly, most companies attempt to do so, and this results in simply wearing out engineering teams with useless activities.

Classic examples include endless meetings and expecting engineers to be involved in non-technical planning sessions, meaningless team events, training sessions and other activities conducted in the guise of team building.

At first, they might be a novelty, giving engineers a new flavor. But eventually, they become a nuisance and the time and energy spent on non-engineering tasks become a burden in addition to the time and energy needed to complete projects.

In all the years I spent working under various managers, I have never seen a manager undertake meaningful and relevant trainings. I must also admit, during my brief stint as a manager, I too failed to carry out worthwhile trainings or growth-oriented activities.

Here, one should look at the matter from the engineering viewpoint. One might think that no engineer wishes to indulge in training sessions, but rather simply complete the necessary tasks and leave. This might be true for a minority. Most engineers appreciate the opportunity to gain meaningful training that enhances their skills. Moreover, the possibility for an engineer to undertake additional side-projects with the blessings and support of the management can result in valuable additions to their resumes.

The art of continuous training in engineering lies in keeping it minimal and ensuring that they are meaningful.

When conducting training courses, except for mandatory ones related to health, safety, and ethics, it is advisable to get the feedback of the engineering team about the education session they would be interested in. In general, companies conduct random training sessions, inviting experts to deliver seminars and sending engineers to conferences and workshops, assuming that these will be of *some* eventual benefit to the team.

Unfortunately, an elementary school mentality where children are taught various subjects in the hope they will build character, cannot be applied to professional life. No professional appreciates their time being wasted, and the same goes for engineers.

In the same way management expects engineers to do their

job such that the management does not have to be involved in engineering issues, engineers too should expect their time not being wasted on irrelevant issues.

Hence, a manager should identify domains relevant to projects that their team can benefit from additional training. The manager should discuss with their team to determine the optimal manner to achieve this improvement—by inviting external experts to conduct workshops, by sending team members to enroll for short-term training provided by universities or external agencies or even inviting a few of the team members to conduct such a workshop.

As an engineer, if one finds oneself with such a manager, seize the opportunity to learn. There are countless examples of engineers who say their reason for staying in a company is due to learning opportunities. There are managers who think spending money and time on training engineers is wasteful and dangerous, as these engineers can use their enhanced skills to find better jobs.

Such thinking is petty. It's worse for the company if employees stagnate and don't leave.

A manager leading a team on skill-building activities is like a prince holding military drills and hunts with his generals. The general's task is much more difficult, of course. A special case arises when a manager was a skilled engineer in the past, in which case they can undertake regular sessions where the team can work on projects completely independent to the company. Such projects can either be chosen by the team, or can be team members showcasing their projects to which the rest of the team can offer their opinions and potentially even join in.

I once met a manager in a software company who set aside Thursday lunches for the team to discuss side-projects, where any team member could talk about their projects with the others offering their advice and feedback.

These sessions resulted in comradery within the team and fostered healthy competition within the team when presenting their side-projects.

The Machiavellian Engineer

But to exercise the intellect the prince should read histories, and study there the actions of illustrious men, to see how they have borne themselves in war, to examine the causes of their victories and defeat, so as to avoid the latter and imitate the former; and above all do as an illustrious man did, who took as an exemplar one who had been praised and famous before him, and whose achievements and deeds he always kept in his mind.

Aforementioned, a manager cannot rely on innate abilities and besides continuously learning new management techniques and processes, the manager must also learn from the past. This learning must not just be with respect to their own past experiences or those of acquaintances and colleagues, but also by reading biographies of industry leaders.

Besides books written by industry leaders themselves, one can find numerous biographies written on industry leaders over the past several decades. Unfortunately, with the explosion of social media, there is a new breed of influencers offering advice and wisdom on a variety of topics.

Though some of the content is genuine and well-researched, a lot of it is shallow and created to expand the following of these influencers in a new index of success. Hence, a manager must carefully choose the content they imbibe. Content that analyzes in detail the work of industry leaders can be a source of great inspiration—besides trying to understand the thought and action that resulted in success or failure.

The same can be said for engineers. Engineers should read the biographies and histories of other illustrious engineers and scientists, not just about what they created, but also how they went about creation. Such reading can be informative and inspiring. When reading about great scientists and researchers, one feels a natural urge to emulate them. In a previous chapter, I wrote about Linus Torvalds and Guido van Rossum as examples of great engineers who inspired me.

Besides them, one can always find biographies of popular scientists such as Nikola Tesla, Richard Feynman, Stephen

Hawking and many others. As the great Albert Einstein once said, *Innovation is 99% perspiration and 1% inspiration.* However, it is my belief that the *1%* of inspiration is crucial, without which no innovation can truly take place.

Continuing the discussion on military preparedness, Machiavelli discusses how the strength of a principality can be judged in a chapter called *Concerning the Way in Which the strength of all principalities Ought to be Measured.*

> And whoever shall fortify his town well, and shall have managed the other concerns of his subjects in the way stated above, and to be often repeated, will never be attacked without great caution, for men are always averse to enterprises where difficulties can be seen, and it will be seen not to be an easy thing to attack one who has his town well-fortified, and is not hated by his people.

Machiavelli wrote that a conqueror will always think twice about attacking a well-fortified principality, giving the example of the free cities of Germany which he believed were the best protected. He presents the cities of Germany as an unusual case, since these cities do not own much land and were not governed by the emperor of Germany.

Yet, their preparedness for an attack was unique during those times, due to which they remained free. A military conquest during those times could last for months if not years, and in that time, a vast number of problems could be faced by the attacker.

The attacker could run out of supplies, money to pay his soldiers or might face a rebellion. If the defending ruler has the goodwill of the people and planned for such an attack, he can withstand the siege if he selectively puts down a few people who might take advantage for their own ends.

He cannot, however, rely on the people to sacrifice their lives to save him. He must provide a sound military solution and count on the goodwill of the people to make sure they do not take advantage and turn him out.

This same preparedness is essential for companies and engineers as well, especially in these modern times when one faces a whole host of uncertainties, either from natural causes or through human greed. While it may be impossible to plan for some of life's most tragic events, it is possible to plan effectively for a vast number of threats.

Engineers should evaluate the strategies employed by a company to withstand threats and produce effective strategies to protect their own positions. Sadly, most engineers are under the impression that all we need to do is capably perform the tasks assigned to us and all will be well.

As previously highlighted, the company one works for is there to make a profit for its owners and the management is responsible for ensuring that targets are met. The company is not ones' family and management may not look after an employee when times are bad, as they have their own positions to look after.

If one's position in the company is no longer something needed to meet targets, one could be made redundant. For most engineers who go through this, it is devastating to realize your skills are no longer valued and you can be cast aside like a used pair of socks. Unfortunately, this is the way of the world, and one needs to be prepared.

In terms of fortifying oneself, a primary requirement is financial planning when one is employed. In recent times, there is a growing call to make financial literacy mandatory for all. Consequently, for an engineer who possesses even mediocre technical skills, becoming financially literate is achievable. Sadly, most engineers neglect this. When they do have jobs, they tend to splurge and spend most of their earnings with little thought to savings or investments. A target should typically be such that a year of employment will result in savings that can sustain one for around three to four months if one were to lose a job. As employment extends to several years, one will naturally see this buffer extending such that even six months or a year of unemployment can be withstood.

The second aspect to fortifying oneself is physical health. Here again, most engineers tend to neglect their health in various ways with poor diets, irregular sleep and lack of exercise. I came across engineers in their twenties who are out of breath when they climb a flight of stairs. Poor health takes its toll during times of stress such as when one loses a job.

In addition to a financial burden, one now has a physical burden as well. As before, it may not be in the interest of an engineer to attain peak physical condition as the effort required is too great and the sacrifices to be made in terms of diet and lifestyle might not be possible. Per contra, basic health is achievable with minor impact to lifestyle. I advise engineers to study what it takes to be healthy.

The next form of fortification is one which most engineers can attain with ease. Augmenting your technical skills. As an engineer, your greatest asset is your skills portfolio. The more diverse and in-depth your skills are, the greater are the chances that you will be retained even if a company is in trouble; and even if one were to lose a job, finding another one will be much easier.

Keep in mind that when companies lay-off engineers during a downturn, their main consideration is—*how critical are the services of the engineer, and how difficult would it be to recruit a replacement at a later date?*

Thus, regularly invest time in learning new skills or improving existing ones. This does not have to be related to the work of the company but can be completely independent of it. Look for interesting projects on the internet and stay connected with innovative people. One of the best ways to improve yourself is through side projects that show passion and commitment.

For any decent engineer, improving on technical skills should be easy, yet few actively do so—partly due to unwillingness to commit to anything technical at the end of the work day. Another part is due to the misinformation spread that what matters for success is not technical skill but rather soft skills. Though soft skills will help one succeed, the core of almost any project is a technical component which needs a skilled engineer.

Therefore, don't neglect technical skills by being under the impression they won't lead to advancement. Apart from advancement, technical skills help you survive recessions when jobs are scarce and companies recruit a selected few with an array of skills.

An engineer should assess how well their company is fortified for the proverbial rainy day when business goes south. When a company mindlessly splurges on useless activities and travel when business is booming, it might give the impression of the company being considerate towards its employees.

In reality, this can be a red flag, as such companies will also indulge in a number of borderline unethical acts such as giving lavish bonuses to its executives or stock buy-backs. These companies will not think twice about firing employees later or cutting critical research and development budgets.

On the other hand, a company which judiciously invests its bounties in expansions that will diversify its goods and services is a far more reliable employer. There are those who believe companies should distribute surpluses to all its employees during these boom times, however, as will be shown in a later chapter, though it might be idealistic, it may not be a wise business decision.

In reference to Machiavelli's two chapters *How Many kinds of Soldiery There Are, and Concerning Mercenaries* and *Concerning Auxiliaries, Mixed Soldiery, and One's Own*, he discusses different types of armed forces: a ruler using mercenaries, auxiliaries and his own army.

Machiavelli was a strong proponent of rulers maintaining a citizen army and was strongly opposed to mercenaries or auxiliaries, giving extensive examples of how rulers who depended on mercenaries or auxiliaries found themselves ruined or having to eventually build their own armies.

Though the underlying exhortation is obvious, his analysis of the effect of mercenaries and auxiliaries can be used in the engineering world as well.

> The chief foundations of all states, new as well as old or composite, are good laws and good arms; and as there cannot be good laws where the state is not well armed, it follows that where they are well armed they have good laws.

The above statement is obscure in the modern world, where in most countries, the military and law enforcement are completely independent entities, and most citizens have no need to bear arms.

There are still a few countries in the world where military service is mandatory, and a few countries with laws guaranteeing their citizens the right to bear arms. However, in medieval Europe, for principalities to be able to defend themselves, it was necessary that the general public receive some military training.

Machiavelli gives examples of France—and before that the Roman empire—which had well-established ordinances concerning arming their citizens, who had a liability to fight in the event of a war. Machiavelli said such principalities were typically those which were well-defended, and those that were secure and could live in peace and prosperity in turn were governed well.

> The mercenary captains are either capable men or they are not; if they are, you cannot trust them, because they always aspire to their own greatness, either by oppressing you, who are their master, or others contrary to your intentions; but if the captain is not skillful, you are ruined in the usual way.

Mercenaries during Machiavelli's times were groups of armed men who could be recruited to fight battles. Machiavelli believed these are either useless or dangerous, for if the mercenaries are incompetent, they are a waste of money and will lose wars or retreat; and if they are competent, they are dangerous, for they can change their allegiances and go against you.

He gave several examples of mercenaries being the root

cause of decay of many states and believes they were responsible for the mess Italy was in during his time. Historically, one of the prime reasons for the extensive use of mercenaries in Italy was the creation of new principalities or republics due to the rebellion of the citizens against their nobles. Therefore, Italy transitioned from an empire into a collection of principalities and republics, some governed by the Church and others by citizens, most of whom being inexperienced in military affairs and were forced to recruit mercenaries.

Using many historical events, Machiavelli proves mercenaries are dangerous, and those few occasions where they were useful, were simply lucky. He describes an atrocious case of back-stabbing where the Milanese enlisted a mercenary by the name of Francesco Sforza to defeat the Venetians, who ended up joining the Venetians instead to destroy the Milanese who recruited him.

Furthermore, Francesco Sforza's father, also a mercenary, was recruited by the Queen of Naples, but instead, abandoned her and she was forced to throw herself at the feet of the King of Aragon for protection. Machiavelli's examples are difficult to read and require parallel research to understand the historic events.

He concludes with this statement:

> *Because from such arms conquests come but slowly, long delayed and inconsiderable, but the losses sudden and portentous.*

Machiavelli describes how mercenaries were useless in Italy because their goal was to sell to the highest bidder while minimizing risk. In many of his descriptions of historical events revolving mercenaries, he spoke of their lukewarm enthusiasm in military campaigns. When they managed to achieve anything, it was due to good fortune rather than sound military tactics.

> *They had, besides this, used every art to lessen fatigue and danger*

to themselves and their soldiers, not killing in the fray, but taking prisoners and liberating without ransom. They did not attack towns at night, nor did the garrisons of the towns attack encampments at night; they did not surround the camp either with stockade or ditch, nor did they campaign in the winter. All these things were permitted by their military rules, and devised by them to avoid, as I have said, both fatigue and dangers; thus they have brought Italy to slavery and contempt.

Auxiliaries were the armies of another ruler called upon by a ruler to fight on his behalf. These could be due to alliances or other pacts between rulers and nobles. Machiavelli believed mercenaries were useless, while auxiliaries are dangerous. If the auxiliaries were to win, they might not leave as they do not do the bidding of the ruler but rather another ruler whom they consider their master. The result is the army that came to assist becomes an occupier.

These arms may be useful and good in themselves, but for him who calls them in they are always disadvantageous; for losing, one is undone, and winning, one is their captive.

Machiavelli gives the example of Pope Julius who appealed to the King of Spain for military assistance in the pope's campaign in Italy. Once again, Machiavelli stresses that it was sheer good fortune alone which caused the Pope not to be at the mercy of the King of Spain, as to everyone's surprise, the armies of the King of Spain were defeated, and eventually the Swiss defeated and drove all the conquerors out.

Due to this rather complicated set of events, Pope Julius found himself free of obligations either to his enemies or to the King of Spain.

As a final example, Machiavelli describes the actions of Cesare Borgia (Duke Valentino), who was described in a previous chapter. Duke Valentino began his military campaign with auxiliaries by enlisting French soldiers. However, he soon found them unreliable, and instead turned to mercenaries, as he feared less from them as compared to the disciplined army of a foreigner (France). Finding mercenaries to be equally unreliable, Duke Valentino destroyed them and eventually relied on his own citizen army.

To draw a parallel to the modern corporate world, one cannot think of an exact equivalent to mercenaries or auxiliaries. However, considering mercenaries and auxiliaries in contrast to a citizen army, one can think of two approaches to engineering— where companies build technology in-house versus those who outsource development in various forms.

In a company or university with good engineering productivity, there will be competent engineers and discipline, for obviously, only this way, will there be a decent result.

In the same way Machiavelli stated that a ruler needs to invest in his own army, an engineering company or university should invest in developing fundamental engineering practices.

Relying on development from sources external might yield temporary results and appear convenient; but in the long run, failure to generate ones' own intellectual property will lead to the company or university's decline.

In my years in academia and industry, sadly, I saw many professors and managers completely neglect the task of building strong teams of engineers.

The reasons are diverse.

Some choose the cost option—*it may be cheaper to import technology than to build it yourself.* Some choose convenience—*why bother building something when it can be bought or ordered?* Some are lazy—*if it is not absolutely necessary, why bother?* Some are averse to risk—*so many things can go wrong in the process; if the chances of failure are moderate or high, is it worth it?*

As engineers, when we hear these things, it is important to recognize them as red flags. Companies or universities encouraging dysfunctional trains of thought will not be places where engineers can grow, and we should find workarounds to chart an independent path.

Before examining solutions, let us examine why this thinking is so dangerous. Importing technology because it is cheaper to do so, might be cost-effective in the short-term. However, importing technology on a continuous basis without understanding the technology itself makes one a slave to the provider of the technology. Eventually building that component in-house is a business decision that involves cost, risk and sourcing, but to not invest in basic engineering related to the technology is unacceptable.

As time progresses, the company stops being an engineering company and turns into a packaging company sourcing components from others—packing and selling them to the end user. This state of affairs also came to academia where students use ready-made components in their experiments and computations.

It speeds up projects and the results look better. Subsequently, the students lose touch with basic engineering and basic computing. Now we have engineers entering the workforce with no knowledge of basic design or programming.

Besides not investing time in understanding basic engineering, another danger lies in not investing time in educating and training one's own engineers.

Though universities and industry in developing countries are now well-established, they still prefer to recruit engineers who return from overseas rather than their own graduates. As a result, they designated their own talent as second-class which fuels brain-drain.

Engineers in developing countries are told either explicitly or implicitly that anyone who remains in the country is deficient, and those who were good all went abroad.

Is there any surprise that developing countries lose talent to the developed world?

Expecting someone who invested a great deal in educating themselves overseas to return to their home countries and rendering their services for a fraction of the compensation and benefits they would receive overseas is to live in a fool's paradise.

There are many countries in the world with very good engineering systems who don't rely on their students being educated abroad. Examples are countries in Europe, Japan, South Korea and a few others, where candidates from overseas are recruited only when a qualified candidate cannot be found locally.

In many developing counties, one comes across numerous excellent engineers who were not offered good positions because preference was given to someone returning from abroad.

Eventually, some will leave for a number of different reasons. Unfortunately, to treat one's own talent as second-class sets a dangerous trend of local talent wanting to leave the country.

And, once they leave, convincing them to return is not easy.

Any engineer being fed the myth that building systems from the scratch is a waste of time, or that training oneself to be fundamentally strong will be useless to one's career, I offer my refutation. Those who utter these words are thinking more of their selfish ends than of engineering quality. When an engineer is swept away with this rhetoric and does not invest in training and self-improvement in the engineering domain, they eventually become disempowered and under the control of the management.

As previously expressed, an engineer being loyal to a company at the cost of engineering will not yield dividends when a business goes south and the company finds ways to cut costs. Managers who encouraged the engineer to focus on non-technical skills will not think twice to fire that engineer.

Consequently, despite the company's approach to engineering, an engineer should not forsake engineering rigor, as even if not immediately applicable, times may change when an opportunity arises in which skill and talent will be recognized.

> *And if it be urged that whoever is armed will act in the same*
> *way, whether mercenary or not, I reply that when arms have to*

> *be resorted to, either by a prince or a republic, then the prince ought to go in person and perform the duty of a captain; the republic has to send its citizens, and when one is sent who does not turn out satisfactorily, it ought to recall him, and when one is worthy, to hold him by the laws so that he does not leave the command. And experience has shown princes and republics, single-handed, making the greatest progress, and mercenaries doing nothing except damage; and it is more difficult to bring a republic, armed with its own arms, under the sway of one of its citizens than it is to bring one armed with foreign arms. Rome and Sparta stood for many ages armed and free. The Switzers are completely armed and free.*

Machiavelli's paragraph quoted above is a powerful reiteration of the use of one's arms over the use of foreign mercenaries. It was a refutation to those who found no difference between mercenaries and a citizen army.

Machiavelli gave numerous examples of mercenaries who failed or had to be disbanded or destroyed to prevent a kingdom's downfall.

On the other hand, rulers taking charge of their armies and maintaining an army of their own subjects achieved the most. He gave examples of kingdoms that were free due to their citizen armies being strong and their subjects being armed. He later also described how rulers who armed their citizens and trained them to use those arms were those who defended their kingdoms well.

This was contrary to the belief that arming one's own citizens came at the risk of an uprising, which Machiavelli maintained was due to the ruler being hated or despised. This will be discussed in a later chapter.

I knew visionary professors and managers who built teams and systems from the ground up and led their team through thick and thin while struggling with the challenges of building from scratch. They were rewarded for their pain by having full control over their products and systems, and their teams—having overcome those challenges—were well-trained and skilled.

Of course, there will always be compromise, where one focuses energies on the greatest challenges at hand. The quality of a great leader is the ability to mix compromise with rigor and answer challenges with solutions leading to the greatest advancements.

> *I wish also to recall to memory an instance from the Old Testament applicable to this subject. David offered himself to Saul to fight with Goliath, the Philistine champion, and, to give him courage, Saul armed him with his own weapons; which David rejected as soon as he had them on his back, saying he could make no use of them, and that he wished to meet the enemy with his sling and his knife. In conclusion, the arms of others either fall from your back, or they weigh you down, or they bind you fast.*

However historically true was the fight between David and Goliath, it offers numerous learning opportunities, and, above, Machiavelli makes an interesting observation. Someone might offer arms for various reasons, either benevolently as Saul did, or maliciously to cause your downfall, but the weapons that are not truly your own will never be wielded with the same skill as you would wield your own.

If one must choose alien weapons, let it not be on the spur of the moment, but from painstaking training and sacrifice, such that these weapons are no longer alien, but as good as the ones that were held before.

In this modern, fast-paced world where technology advances rapidly, one should not look inward and hold on to only that which one possessed.

However, if the process of adopting another technology is done without rigorous training and analysis, those weapons would be the same as those Saul handed to David just before the battle.

Therefore, a sign of a good engineering company is one that embraces new technology but has a fundamental understanding to not only use it successfully, but also modify and use it in the best possible way.

> *After these came all the other captains who till now have directed the arms of Italy; and the end of all their valour has been, that she has been overrun by Charles, robbed by Louis, ravaged by Ferdinand, and insulted by the Switzers ... All these things were permitted by their military rules, and devised by them to avoid, as I have said, both fatigue and dangers; thus they have brought Italy to slavery and contempt.*

Machiavelli described in detail how mercenaries and auxiliaries led to the downfall of Italy. What started as an unfortunate necessity, where those who were new to government and had to resort to mercenaries and auxiliaries, then sadly became a convenience. From that convenience sprang decay.

The principalities of Italy maintained the status quo between them using these weapons of convenience, but in the face of a foreign invader, caused the downfall of Italy. Italy faced a number of conquests, initially from France and then from Spain.

Unable to defend itself, it ended up changing invaders, when one (Spain) drove the other (France) out.

> *I conclude, therefore, that no principality is secure without having its own forces; on the contrary, it is entirely dependent on good fortune, not having the valour which in adversity would defend it. And it has always been the opinion and judgment of wise men that nothing can be so uncertain or unstable as fame or power not founded on its own strength.*

After all these discussions, the question again is, *what must an engineer do?* As mentioned earlier, most engineers cannot change anything within a company. Managers make decisions without consulting engineers.

However, when in a company or a university where the management is content with superficial work that brings them temporary and short-term gains, an engineer should acknowledge that this is unsustainable and embracing such a philosophy will cause significant damage to their career.

While it is essential to fulfil one's obligations to maintain a position in the company, an engineer should not limit oneself to the mere tasks of the company. This was stressed in the previous chapters as well. However, the difference is it might be difficult to understand the superficial nature of work is detrimental, since in the short-term one might see some gains.

Refer to Machiavelli's interpretation of the downfall of Italy where the Italian republics and principalities maintained a semblance of peace in the absence of external factors but were overwhelmed when France invaded.

Working for a company that emphasizes quick wins might seem like progress under normal times, but, in the event of an economic downturn or in the face of competition from an aggressive rival, the smooth ride unravels.

To re-visit earlier reference, when a company is in trouble, employees are left to fend for themselves with junior and contract employees being the first to face the axe. Becoming accustomed to the trivial nature of company work, engineers can lose their competitive edge and if the company folds, get left out in the cold.

> But the scanty wisdom of man, on entering into an affair which looks well at first, cannot discern the poison that is hidden in it, as I have said above of hectic fevers. Therefore, if he who rules a principality cannot recognize evils until they are upon him, he is not truly wise; and this insight is given to few.

Hence my suggested advice to young engineers who find themselves in such situations where their company or university shows little regard for them as valued assets, is to not seek their approval or tie their opinions to your feelings of self-worth.

In the end, managers and professors will face the consequences of their actions. In the meantime, stay focused on building skills, stay connected with other motivated and talented people and engage in interesting, fruitful endeavors as described in the previous chapters.

As time passes, the time and energy you and your network invested will bear fruit and the dinosaurs who presided over decay will be too weak to obstruct you.

This chapter combined several chapters of *The Prince* dealing with military preparedness which Machiavelli promoted. The relevant chapters of *The Prince* are a wonderful read with many references to historical events followed by in-depth analysis. This analysis sets Machiavelli's writings apart from others who wrote similar works.

The analogy between military preparedness and engineering is obvious. Using Machiavelli's analysis in a coherent manner requires repeated reading and changing your behaviour. After spending many years as a graduate student and later working in industry, I arrived at similar conclusions with regard to engineering training.

As stated in the previous chapters, the primary goal of any company is to make a profit for the owners and shareholders. As long as the company makes a profit, management would like to continue a strategy of *business as usual*, which focuses on profitable processes.

On the other hand, for an engineer, the primary goal should be securing positions and increasing their employability. By continuously training and expanding skills, engineers find their positions less precarious when a business goes south.

Companies retain those who have a wider array of skills.

Though it's obvious that continuous training is imperative, engineers tend to neglect it, and instead immerse themselves in their companies. Such dedication yield fruits if the company is progressive, technically competitive and values and rewarded the efforts of engineers.

However, in many cases, engineers waste their time in petty or unnecessary tasks dumped on them in the guise of *corporate culture*.

In his writings, Machiavelli described how important it was for rulers to invest in their own citizen armies and to fortify their kingdoms. He described many historical events where dependance on mercenaries proved to be the downfall of rulers. I used Machiavelli's analysis to describe the importance of core engineering acumen for both companies and engineers.

However, this problem is acute for engineers. When they believe the propaganda of companies and neglect engineering training, they put their livelihoods at risk. In the past chapters, I describe the benefits of having a side-project and how these side-projects can be nurtured and sustained under different circumstances. In the context of engineering training, side-projects provide an excellent sandbox for expanding and honing skills.

This chapter is named *Technical Weapons* for the simple reason that in these modern times, technical skills are formidable weapons. Engineers benefit when they continuously invest in training, but sadly, only a few do.

One reason is the false information provided to them. Another part is the artificial pressure created by company managements to ensure that engineers do not find the time for training or side-projects.

This chapter and the previous chapters described ways engineers can find the time to invest in training or indulge in side-projects.

Though these methods appear deceitful, they are no less deceitful than management strategies focused only on augmenting their own profits.

The Deception of Perfection

IN THIS CHAPTER, I combine thoughts from three chapters from Machiavelli's writings—*Concerning Things for Which Men, and Especially Princes, are Praised or Blamed*, *Concerning Liberality and Meanness*, *That One Should Avoid Being Despised and Hated*, *Are Fortresses, and Many Other Things to Which Princes Often Resort, Advantageous or Hurtful?*

In these chapters, Machiavelli writes about the various strategies that a ruler can use, which might appear at first to be qualities detestable in a ruler.

However, Machiavelli argues that for a ruler to appear to be perfect is unrealistic and describes many qualities which though might appear as defects in a ruler's personality, can be greatly beneficial if used appropriately.

As with the other chapters, I describe how these qualities can be translated to the modern corporate world and used both by managements and engineers.

> *But, it being my intention to write a thing which shall be useful to him who apprehends it, it appears to me more appropriate to follow up the real truth of the matter than the imagination of it;*

for many have pictured republics and principalities which in fact have never been known or seen, because how one lives is so far distant from how one ought to live, that he who neglects what is done for what ought to be done, sooner effects his ruin than his preservation; for a man who wishes to act entirely up to his professions of virtue soon meets with what destroys him among so much that is evil.

Machiavelli begins this collection of chapters by stating that one can imagine an ideal kingdom with an ideal ruler, ideal nobles and ideal subjects, and with assumption, build a fairy tale suitable for children.

Reality, though, is vastly different from the ideal. A ruler governing in an ideal manner will—almost always—be ruined. Machiavelli wrote many controversial chapters on the necessity for a ruler to compromise with honesty, integrity and reverence.

He believed a ruler preserving his throne and his kingdom in the midst of numerous threats from both within and without, could not be purely virtuous, for his evil opponents would not hesitate to use every means at their disposal to destroy him.

Machiavelli wrote about things that were widely known and practiced by many. What makes his philosophy remarkable is the detail in which he presented his analysis with examples of the past and contemporary rulers of his time. His teachings are still used by modern politicians and business leaders. One can find many examples in politics and the corporate world.

Hence it is necessary for a prince wishing to hold his own to know how to do wrong, and to make use of it or not according to necessity.

A part of this was discussed before in two parts—of how kingdoms were attained by wicked means, and how cruelty is necessary for ensuring loyalty because it is better for a ruler to be feared than loved—if not both.

Machiavelli said that using wicked means should be the last

resort and should cease once the conditions change, rather than multiply with time. Machiavelli also believed cruelty practiced by a ruler might be essential to put down dissenters, and a reputation for being cruel was not necessarily a bad one as long as the ruler is not hated or despised.

Machiavelli added a word of caution: that a ruler choosing to be cruel or wicked must do so as the need arose, and for the safety and security of the kingdom, and not out of wanton malice, whim or fancy. He stated in this context of wickedness and cruelty, a ruler must strive to avoid being despised or hated, either by the nobles or by the people.

This concept might seem vague. In this chapter, I describe in detail Machiavelli's philosophy about how a ruler should avoid being hated or despised, and how so it would be applicable for the engineering world.

> And again, he need not make himself uneasy at incurring a reproach for those vices without which the state can only be saved with difficulty, for if everything is considered carefully, it will be found that something which looks like virtue, if followed, would be his ruin; whilst something else, which looks like vice, yet followed brings him security and prosperity.

No human is fully virtuous, and one should not expect unconditional virtue from a ruler. A ruler must ensure the safety and security of his kingdom, and for this to be guaranteed, it is not only virtue that matters, but vices as well.

A ruler should be good as well as evil, to be just as well as cruel—he must be willing to do everything necessary to augment his rule. A ruler must be willing to do wrong in manner that does not weaken his position.

Accordingly, a ruler cannot avoid reproach for certain vices necessary for the security of the state. The same can be said for companies and employees. Companies use various methods and schemes that even if not illegal, are unethical, and as long as the company flourishes, these methods do not attract condemnation. This strategy is scandalous when we expect our leaders to be saints or paragons but know that no leader can afford to always be just.

To begin this discussion on the qualities of a ruler and how they deviate from an ideal image of a ruler, Machiavelli speaks of liberality and meanness—he speaks of liberality in rulers in an interesting manner easily translatable to the modern corporate world.

The opposite of liberality is meanness or miserliness, where the ruler makes every effort to save money.

> *Therefore, a prince, not being able to exercise this virtue of liberality in such a way that it is recognized, except to his cost, if he is wise he ought not to fear the reputation of being mean, for in time he will come to be more considered than if liberal, seeing that with his economy his revenues are enough, that he can defend himself against all attacks, and is able to engage in enterprises without burdening his people; thus it comes to pass that he exercises liberality towards all from whom he does not take, who are numberless, and meanness towards those to whom he does not give, who are few.*

The first part of the quote above deals with the ideal notion of being charitable in secret versus being magnanimous in public. Though being charitable in secret might be praised, the giver gets no recognition of being magnanimous.

On the other hand, for a giver to be praised for being magnanimous, the gift must be significant. For this reason, Machiavelli stated that a ruler who was liberal and made a show of magnificence ran the risk of depleting his wealth to get the reputation of being generous.

However, such a ruler then would have to tax the citizens or worse, confiscate their property if he wished to continue being liberal and generous. A ruler who was frugal would be in a good state to undertake military expeditions and conquests without burdening his citizens. The only ones offended would be the nobles who, being a few in number, would not pose a serious threat as long as the ruler took care of his personal defenses.

> *Therefore it is wiser to have a reputation for meanness which brings reproach without hatred, than to be compelled through seeking a reputation for liberality to incur a name for rapacity which begets reproach with hatred ... A prince, therefore, provided that he has not to rob his subjects, that he can defend himself, that he does not become poor and abject, that he is not forced to become rapacious, ought to hold of little account a reputation for being mean, for it is one of those vices which will enable him to govern.*

Machiavelli was always adamant that a ruler should avoid being hated. When the hatred spreads, conspiracies grow and malcontents can grow strong enough to pose the threat of a rebellion. As long as meanness is practiced in a manner that does not result in the subjects or the nobles hating the ruler, a reputation for being mean will result in reproach from those who might have expected generosity.

Unfortunately, the exercise of liberality to avoid reproach was foolish and dangerous, as a subsequent inevitable reputation for greed to sustain the liberality would breed a much harsher reproach along with the hatred of those who had been injured.

Machiavelli presented a few examples of rulers such as Pope Julius the Second or the King of Spain who fought wars with their own resources.

> *Either you are a prince in fact, or in a way to become one. In the first case this liberality is dangerous, in the second it is very necessary to be considered liberal ... And if anyone should reply: Many have been princes, and have done great things with armies, who have been considered very liberal, I reply: Either a prince spends that which is his own or his subjects' or else that of others. In the first case he ought to be sparing, in the second he ought*

not to neglect any opportunity for liberality. And to the prince
who goes forth with his army, supporting it by pillage, sack, and
extortion, handling that which belongs to others, this liberality
is necessary, otherwise he would not be followed by soldiers.

Machiavelli believed liberality could only be justified under two circumstances—when a new prince was consolidating his rule over his kingdom, or when a ruler was conquering another kingdom in a military expedition.

In the first case, the new ruler must gain the goodwill of the people and the nobles, and for this must appear to be generous, so that they believe that such generosity will continue to follow during his rule.

In the second case, the ruler must allow his generals and soldiers to pillage and seize the property of those who are being conquered to maintain the goodwill of his soldiers and avoid a rebellion. Once he ascended to power or consolidated his power following the military conquest, he must curtail his expenses and follow a frugal method of governing.

Thus, being liberal must only be a temporary strategy to gain the goodwill of those without which one cannot either stay in power or continue an expedition.

In all companies, whether family businesses or mega corporations, management or the owners are at a perennial conflict of whether they should appear to be liberal or frugal. Conversely, employees prefer that management make every attempt to be generous with their compensation and benefits, while also ensuring stable and continuous employment for which it is of course imperative that the company make sufficient profits.

It is obvious there are conflicts in the logic above—a company cannot be lavish in compensation and benefits without losing a significant share of the profits. With the retained earnings of the company limited, this provides a thin safety net for the proverbial rainy days when a business revenue might dwindle.

How a company should be run and which financial model it

should adopt is not at all the objective of this book. As earlier expressed, the way a company is run will have a direct impact on the well-being of the employees and herewith, an employee should judge whether the economic policies of the company are sound.

Machiavelli's philosophy on liberality and meanness can be applied with surprisingly few modifications in diverse circumstances. As someone who worked in various companies ranging from startups to mega companies, I applied Machiavelli's philosophies in many circumstances, and they were always good learning experiences.

We who worked in companies have had mean bosses who invest very little in the well-being of teams, though they had at their disposal sufficient funds to make the lives of their team members much better. Unfortunately, we view our managers and professors as parental-figures and the unrealistic expectations of our companies and leaders lead to frustration with actions that are normal.

It is important to remember that while it is reasonable to expect the company to bear essential work expenses, it is far-fetched to expect a company to finance expenses that do not have an immediate impact on productivity. As an example, it is reasonable for employees to expect a company to provide facilities necessary to work. If management is unwilling to provide necessary facilities to perform work, they will run the company to the ground. Management unwilling to deem an expense necessary when it is, is incompetent and such incompetency will spill over to other aspects of management.

On the other hand, if management declines expenditure an employee is enthusiastic about, but not truly essential, is not necessarily a sign of poor management, but might just be a sign of thrift. Examples could be an unwillingness to sponsor travel requests or purchase expensive equipment that may not be fully utilized.

Instead of getting frustrated at requests being rejected, an employee should stop and ask the question—*is the denied request absolutely essential?* If the request represents something that would be good to have or a good to do, but without tangible and concrete business value, the behavior of management need not be perceived as negative.

Instead, the employee should compare the behavior of the management to various expenditures and notice where the management does approve spending and where it declines. If management decisions are relatively sound and expenses are approved when deemed necessary, such a management need not be a bad one, and in the long run, the company might flourish and grow.

In these modern times, funding cuts are common, and many critics of the corporate world are quick to condemn any cut in funding. As expressed, it is impossible to throttle the inclination of the management to maximize the profits of the company, as the compensation and bonuses of the management are tightly coupled with the profits that the company posts.

All employees can do is adjust their investment in a company according to the wisdom shown by the management in running the company. In this manner, to distinguish a reasonable cut or diversion in funding from a completely unscientific or whimsical decision, one needs to be objective.

While working in companies, employees tend to fall in factions, where one faction will hold the management's decisions as the words of the messiah, while another might think of every management decision originating in pure evil. Avoiding stark polarization is hard, and often comes from experience, which in turn is the result of poor decisions and mistakes.

There are many cases where liberality or the outward show of liberality and grandeur is disastrous to employees. In recent times, one hears of companies terminating hundreds of employees at a time, and in some cases with no notice period or advance notification allowing them to plan the next step of their career.

In many cases, these companies expanded rapidly, recruiting

a vast number of employees while business was good with little consideration as to whether the growth was sustainable or feasible.

Companies give the impression of rapid expansion to get a bump in the stock price or the valuation of the company, which benefits primarily the executives and investors. When business revenue declines, these companies will quickly turn around and let go employees who had been recruited, treating them like commodities. Some companies will go an extra mile in using the tumble in the stock price of the company to buy back stocks which were sold during the boom time.

Company management was either short-sighted and living in the moment, or greedy and selfish. Rather than retaining business revenue, they spent to meet the additional business demand, or filled their pockets with fat bonuses and stock options. It would have been prudent to let the increased demand result in additional wait times for the product or services to be supplied.

To project an impression of success and prosperity to shareholders and employees, the company consumes surplus profits. When forced to let go employees, it sends the signal to shareholders and employees that management is fickle and unreliable by being turbulent and unpredictable—exactly the opposite signal it intends to send.

As Machiavelli showed, a big-spending ruler will eventually resort to taxing his citizens or confiscating their property. In the attempt to project itself as a unicorn or a super successful company on an incredible growth trajectory, the company is forced to admit the exact reverse. A company treating its employees as objects that can be discarded when inconvenient will rarely have a loyal workforce and will resort to nefarious means to extract work from them.

As Machiavelli stated, instead of a mild reproach for frugality from employees, the company earns hatred or contempt. Unlike in medieval times when there was the risk of an armed uprising, these companies struggle to retain employees instead of facing real rebellion.

Another example of how liberality of management can cause the resentment of the employees is how surplus profits are distributed. When business booms and a company reaps profit, it pays out lavish bonuses to its executives and preferred shareholders and a minor bonus to the rest of the employees.

Often notably, the company sets aside a major portion of the profits to buy back shares to retain greater control of the company and minimize dividend payouts to external shareholders.

This is a major bone of contention for those whose political beliefs lean towards the left, and as in these cases, the company profits largely benefit the top 1%. The socialist belief favors distributing profits equally among employees. Except for the case of cooperatives, most engineering companies are corporations or mere family businesses, where the profits will rarely reach the employees.

From the perspective of how a company should be ideally governed, the profits of a company should be reinvested into processes that generate more profits. More profits can be generated by broadly two paths—expansion of products and services and improvement of existing systems.

I have already spoken about expansion. Such expansion usually involves recruitment. To expand one's products and services, one needs to invest in a larger workforce. Investing only in expanding the workforce is unfortunately the only method used by many companies, and this brings to the limelight mass layoffs that almost always follow the expansion.

Few companies proactively set aside a portion of profits toward improving existing systems. As employees, and particularly as engineers, look for signs that a company invests in improving itself, and more importantly, in an incremental manner.

This is an excellent example of where a company's management appears to be mean, as employees may not see an increase in their compensation or benefits, nor may perceive the company as one that is on a rapid ascent.

A company that retains profits towards improving its systems is a rare one, and if one finds signs of such investment, one should see those as signs of progressive management. Despite what might appear to be wage stagnation, jobs in such companies are stable and yield satisfaction. A management that is circumspect about new recruitment will value existing employees to a greater extent.

Unfortunately, it is rare, and even when companies invest in research, it is only when the company realizes it may end up being obsolete in the face of its competitors.

Apart from forming a perspective on the nature of management and how it impacts them, employees can practice liberality and meanness in turn. Here, significantly modify Machiavelli's philosophy as meanness or liberality on the part of the management which is typically related to how they use the profits of the company.

In the case of employees, all an employee has to offer is their skills and energy. As a result, to be liberal or mean, an employee should decide how they intend to use their skills and energy in the company.

This was discussed before.

Though a company expects 100% commitment and dedication from every employee, it's foolish for an employee to simply give what is asked. In the same manner, company management would reject socialist ideas where profits are shared equally between all employees.

To be liberal with one's skills and energy means the employee devotes themselves completely to the company. Such liberality is expected and even demanded when employees work long hours to the point of physical and mental burnout. The company benefits. High performing companies have talented employees giving their very best.

However, one needs to ask the question—*what does one get in return? A Top Performer badge?* Only in a few companies is dedication well-compensated—either through bonuses or promotions. In such cases, the talented and hard-working are successful. Unfortunately, in most companies, hard work is not well rewarded, and those who do so are liberal at the expense of their personal lives and health. Employees can become bitter and resentful and lose interest in their jobs, which might further jeopardize their positions.

Once again, this is an example of liberality resulting in the exact opposite of what was intended.

As Machiavelli stated, there are times when it is essential for an employee to be liberal. When starting at a new company, one needs to make a positive impression, and in such a case, similar to a new prince gaining control of a kingdom by winning over the nobles and citizens through liberality, the new employee should work long hours and gain the confidence of the management.

However, in the same way that a prince who remains liberal upon ascending the throne ends up destroying his assets, employees who continue to work incredibly long hours even after establishing themselves in a company risk burning themselves out.

In the same way a ruler needs generosity to ensure peace and goodwill in the kingdom, employees should be generous with their efforts so management has confidence in them.

In these modern times where employees face continuous appraisals, we're compelled to work out of fear of being punished or reprimanded. The scale of punishment is totally dependent on the company.

In some cases a company works in an extremely competitive domain and underperforming employees face termination with short notice. In most cases it takes long periods of incompetency before one faces the threat of termination. Employees who cut back on their work output might face reproach from their managers, but if this reproach is mild and appears to be *more bark than bite*, one need not fear company management.

Machiavelli said a frugal ruler who grows his wealth can later raise an army for a military conquest or for defending his kingdom from an attack without burdening his citizens with additional taxes.

In the same manner, an employee who dilutes investment in the company can utilize excess time and energy to undertake training or indulge in side-projects or independent business activities. This in turn will increase their employability or provide alternate career options. In most cases, employees perform menial and mundane tasks because the nature of a profitable business involves repeatability. For employees, burning themselves out with mundane work to gain the favor of the management is wasteful.

Machiavelli believed a ruler should not be concerned about being branded cruel, wicked, mean, or deceitful as long as he is not hated and despised. One would assume any of those qualities will cause a ruler to be hated and despised, but Machiavelli believed that a mere reproach for these qualities would not border on hatred or contempt.

On the other hand, a reckless ruler who behaves in a manner causing the citizens and the nobles to hate him will have to resort to an inhuman degree of cruelty or eventually lose the kingdom.

Machiavelli wrote an entire chapter about how a ruler should avoid being hated or despised—with many detailed examples related to Roman emperors. As with most of his analysis of history, it is complicated and needs several readings to fully understand every nuance.

> ...the prince must consider, as has been in part said before, how to avoid those things which will make him hated or contemptible; and as often as he shall have succeeded he will have fulfilled his part, and he need not fear any danger in other reproaches.

In the modern corporate world, the possibility of a violent takeover of a company is minimal. Intrigues at the executive level will not be discussed because this book is targeted at engineers and employees.

One might be quick to dismiss Machiavelli's philosophy for this reason.

However, one needs to remember that intrinsic human nature plays a major role in employees leaving a company. Many are motivated by hatred or disgust when exiting a company and satisfaction or contentment if not affection and admiration to continue in their jobs.

A poorly managed company faces high attrition with employees leaving after short periods of employment. Such attrition eventually becomes externally visible and prospective employees whom the company wished to recruit become hesitant to come onboard.

> *It makes him hated above all things, as I have said, to be rapacious, and to be a violator of the property and women of his subjects, from both of which he must abstain. And when neither their property nor their honor is touched, the majority of men live content, and he has only to contend with the ambition of a few, whom he can curb with ease in many ways.*

It is hard to imagine a parallel between the life of a ruler and that of a manager or professor. Unless in the case of hazardous industries such as mines where labor laws are weakly enforced and one hears tales of terrible mistreatment of workers, one would usually not expect a manager to indulge in acts that a ruler during Machiavelli's times would have.

This makes us ask the question—*when would a manager or a professor be considered rapacious and equivalent to a violator?* Many who survived toxic workplaces will affirm that a manager responsible for creating a toxic atmosphere is often hated.

A manager who verbally abuses employees and makes the workplace detestable is someone often depicted in television and cinema. Sadly, many of us have experience with these types of managers during our professional lives.

Such managers make it agonizing for an employee to come

to work, eventually hating their jobs and the manager.

In previous chapters, I talked about strategies used by company management to pressurize, suppress, or even harass employees such as deliberately fostering nepotism. Though these strategies could cause employees to lose interest in their jobs to the extent that they wish to leave. A workplace where employees are in constant fear is one which is usually the most unstable.

In the chapter *The Urge to Destroy*, I described how this strategy of instilling fear in employees is practiced by many companies, particularly the large and well-established ones.

On the other hand, if employees can work in relative peace, the other strategies employed by the management need not force them out if management has other beneficiary programs in place.

When faced with a manager who behaves like a gangster with little regard for their employees, it is best to find another job.

> *It makes him contemptible to be considered fickle, frivolous, effeminate, mean-spirited, irresolute, from all of which a prince should guard himself as from a rock; and he should endeavour to show in his actions greatness, courage, gravity, and fortitude; and in his private dealings with his subjects let him show that his judgments are irrevocable, and maintain himself in such reputation that no one can hope either to deceive him or to get round him.*

To hate one's manager or professor is extreme and should only happen when the manager or professor behaves egregiously. There are many reasons why one would find one's manager or professor contemptible or to use a simple word: cheap.

Many qualities Machiavelli spoke of can be found in many managers or professors. As a matter of fact, a vast number of managers and professors fall into this category for various reasons, primarily because—unlike a medieval ruler—most managers and professors attain their positions through a process of natural growth, rather than a major historical event, and therefore, are in many cases, of mediocre caliber with weak leadership qualities.

Though almost everyone has a few detractors and even a good manager will have someone finding them contemptible. To possess leadership qualities commanding the respect of the majority is still rare.

A fickle manager or professor indulges in favoritism or has opinions that change often and even worse uses these opinions as a foundation for making decisions. Such leaders are generally shallow people craving unearned respect and turn to mediocre employees willing to shower them with empty praise or simply agree with all their unfounded opinions.

This was described in detail in past chapters. The success of such managers is varied since companies differ vastly. In well-established companies without a great deal of competition, these managers tend to survive and thrive for the simple reason that there are higher-ups of such a nature as well.

In companies operating in a competitive domain or in smaller companies, many such managers fail because decisions made based on wavering opinions lead to failure.

The margin of error in such companies is low.

For an engineer, such a manager is either a blessing or a source of frustration. Due to the favoritism that is usually the case when there is such a manager, the quality of work usually is either mediocre or downright poor.

Good work is not acknowledged and mediocrity gets placed on a pedestal, all to the frustration of the skilled engineer. In such cases, it is advisable for the skilled engineer to scale back their work output, rather than waste time and energy.

Such a manager can be a blessing in disguise for engineers who have fruitful side projects. There have been many cases where engineers jump started their side projects and advanced them while working under such management. All one needs to appease such a manager is a few empty words of praise.

A manager or professor who is frivolous is one who is not driven towards meaningful achievements or rather is focused on promotions or winning awards that have dubious standards.

Companies and universities are full of people who decorate their walls with certificates, awards and trophies and are driven towards collecting more just as a child collects toys.

Such ambition in a leader seems commendable at first glance, but rarely leads to meaningful growth. As we are all aware, a vast number of awards are the result of lobbying and campaigning and to be the recipient usually means either the recipient campaigned with the committee bestowing the award or another influential person did so on the recipient's behalf.

Engineers are quick to assume it's beneficial to work under such a manager, as the manager's success or winnings eventually trickle down to the engineer. Sadly, from experience, I advise against such a belief. For the engineer, such a manager or professor poses a significant risk.

Such frivolous managers are likely to steal the engineer's work and not give them credit.

I witnessed this countless times and, on a few occasions, was at the receiving end as well. Engineers should differentiate between a manager who genuinely wants to achieve a commendable goal versus those who chase their own vain pursuits. It is normal to think of any ambition as worthy, but when a leader pursues only goals that benefit themselves, supporting them will usually end in frustration.

On the topic of mean-spiritedness, there is much confusion. As Machiavelli stated, cruelty is a characteristic any leader will need at some time. In the same way a ruler may need to take the life of someone who might be a threat, a manager or professor might need to willingly hurt someone.

In such cases, examine the cruelty very carefully. A calculated and well-planned act that causes damage in a targeted manner with minimal other repercussions doesn't necessarily mean the leader is a bad one. When a leader indulges in acts that damage employees without any noticeable gain but are acts of pride or ego, this is a sign of mean-spiritedness. A mean-spirited manager or professor will sooner or later attack an engineer for no justifiable reason.

These leaders lack genuine direction and objectives and seek pleasure in wanton acts of destruction to give themselves a feeling of worth. They do so either believing they will never be challenged, or they have the support of higher-ups.

Whether they survive or not is a question of their relationship with higher-ups. In any case, engineers devoting themselves to such leaders will result in frustration.

Irresolution occurs when a manager or professor is unable to make sound decisions and avoids them or indulges in knee-jerk reactions. Unfortunately, most managers and professors are either always irresolute or show irresolution at various times. To determine if a manager is irresolute, engineers should carefully examine how a manager makes decisions. *Does the manager examine the matter at hand in detail, collecting information from various sources, or do they listen to a few biased opinions? Is the manager willing to organize major changes in a planned and deliberate manner, or is satisfied with a few cosmetic changes, or even worse with letting things take their course even if obviously the end may be a disaster?*

Engineers should ask these questions as they study their managers.

What should one expect of their managers and professors? As Machiavelli states: *greatness, courage, gravity, and fortitude* and *sound and irrevocable judgement.*

In terms of greatness, differentiate between a leader who wishes to achieve something truly valuable for the greater good versus someone who is after vanity and petty goals. To achieve something valuable to the greater good is something that will at least advance the organization or university if not be of benefit to people in general.

Whatever one may say about the pioneers in different fields as human beings, they left humanity in a much better position. Modern life as we know it—smart phones, computers, cars, airplanes and many others, would be unimaginable if it were not for the vision of a few great people. They may have won awards or become unimaginably rich due to the process, but almost always they strove to achieve something that would change lives for the better.

Very few will find such leaders, but if one finds a manager or professor who genuinely wishes to achieve something that will bring about significant change, such a leader is definitely worth an investment in time and energy.

Of courage and fortitude, it is a tricky to judge a leader. These qualities are exhibited when under pressure or during a setback. A manager who stays calm and focused during times of high pressure is one who can lead a team through thick and thin.

Most managers resort to procedure and routine, but when these fail under adverse circumstances, they buckle and crack.

Dealing with setbacks and failure is an extremely important leadership trait. Leaders assigning blame or resorting to disciplinary action will never assume true command. Leaders who remain calm, examine their options, introspect on mistakes, and devise well-planned future courses of action are rare and genuine.

One expects a leader to lead. It is important to remember that to lead effectively, one must continuously learn. Leaders who constantly seek to understand matters in-depth, who urge their team to exhibit similar thoroughness, and use the knowledge gained while making decisions are leaders who are the least likely to fail.

When leaders who make sound decisions stand by their decisions and hold themselves accountable, such determination is usually the sign of greatness. On the other hand, managers who rely on repetition and lack the depth to understand the nuances of a situation, who do not reflect on past actions and outcomes, are frustrating to work with.

At great length, Machiavelli described the lives of some Roman emperors and how their nature caused their downfall. Though a very detailed description, a few important learnings can be extracted. The Roman emperors contended with three sources of power—the citizens, the senate (nobles) and the army. In most other kingdoms, the army was not an independent source of power. However, Rome differed from them. Due to the expansion of Rome and the large number of provinces governed largely by the army, it was an incredibly powerful entity an emperor had to maintain friendly relations with. This created a conflict for the emperor. The bloodthirsty nature of the army wished to repress the citizens, while the citizens wanted to live in freedom from terror.

The ability to negotiate between these axes of powers eventually decided the emperors' fate.

> *Which course was necessary, because, as princes cannot help being hated by someone, they ought, in the first place, to avoid being hated by everyone, and when they cannot compass this, they ought to endeavour with the utmost diligence to avoid the hatred of the most powerful.*

With the above statement, Machiavelli stressed that Roman emperors who avoided the hatred of the army were those who ruled successfully. In the detailed analysis of Roman emperors presented by Machiavelli, a few were men of peace such as Marcus, Pertinax and Alexander. They met with sad endings, except for Marcus who inherited the throne.

This is in keeping with Machiavelli's earlier discussion on hereditary principalities accustomed to living under the rule of a dynasty. It would take sustained incompetency and misrule to destroy a hereditary principality as those who have been living content under one ruler will continue to do so under his heirs.

As for the other Roman emperors who were men of peace, they were murdered by the soldiers who preferred a warlike emperor who allowed them inflict cruelty on the people.

Machiavelli goes on to describe the fate of the warlike emperors such as Commodus, Severus, Antoninus Caracalla and Maximinus. These were cruel and terrible men who gave their soldiers freedom to commit atrocities.

One would think this cruel nature would be sufficient to guarantee their rule, but except for Severus, they were all murdered by the army. Severus was cruel and oppressed the people, but also, he was a man of such valor and ingenuity, that the people dared not rise against him, and the soldiers respected and admired him.

Severus maintained his throne not just by bravery, but also by trickery and deceit, which Machiavelli said was extremely important in a ruler.

As for Antoninus and Maximinus, despite being cruel and warlike men, their cruelty was unbounded and the countless murders they committed caused them to be hated not just by the people, but also by the nobles and sections of the army.

Eventually, these men were murdered by the army, who though had supported and benefited from the emperor's cruelty, eventually were on guard lest that cruelty be turned against them.

Though Commodus was also cruel and warlike, he indulged in what was during those times considered vile acts for an emperor, such as competing with gladiators in the theater. This made him held in contempt by the army in addition to being hated by the people. His fate was that he was eventually murdered by the army.

One might think these discussions of ancient Rome are largely useless in these modern times. Unfortunately, the modern corporate world resembles the power structure of the ancient Roman empire.

A manager must contend with the employees, upper management to whom they are answerable, and keep the office bullies satisfied. Like ancient Rome, employees have the least power and some managers care little about them.

Typically, managers care the most whether the upper management has confidence in them and appease office bullies who, if not given an opportunity to harass and belittle coworkers, can threaten the position of the manager.

Most engineers can attest to the presence of such office

bullies who enjoy their pursuits with little impunity. One wonders how they are free to operate when the management is quick to initiate disciplinary action against other employees who commit even minor transgressions against company policy.

In the similar manner to which the Roman emperors found it necessary and even beneficial to allow the army to terrorize the common people, company managements find these office bullies a convenient unofficial tool to repress workers and keep them in line.

In a similar manner to an army proving to be the undoing of a Roman emperor, office bullies can lead to the downfall of a manager. By observing the relationships between the management and office bullies, employees can learn a lot about the company and the future potential it offers.

As Machiavelli stated, except in the case of hereditary rule where a ruler faces far fewer challenges to the throne, in most other cases, a ruler needs to gratify various entities by allowing them to commit wrongdoings, or they might be a direct threat to the throne. Likewise, in some very well-established family businesses, members of the founding family hold key positions in the company and outsiders have little or no say in operations.

In most other corporations, with various power struggles and factions, management won't have secure positions and will allow certain groups of employees unusual degrees of freedom. Hence, one will find these bullies, as the management is challenged to keep all employees in their place, so they will prefer to gain the support of these bullies in exchange for turning a blind eye to their bad behavior.

The relationship between managers and office bullies can be like the ways Roman emperors dealt with their armies. A rare case is that of a manager possessing excellent organizational abilities who is by nature commanding and authoritative.

Such a manager allows employees to work normally, ensuring that projects progress and conclude, while keeping a lid on the activities of bullies. In such cases, office bullies are minor irritants to be taken in stride the same way one would deal with the traffic one finds while commuting.

In all my years in industry and academia, I found only one professor who was such a leader with vast technical knowledge besides being an efficient organizer, such that students enjoyed working with him, and those few inclined to bully others were rarely able to do so since no one dared question the authority of the professor.

Another case is that of a reasonable competent manager who possesses either organizational or technical skills necessary for the smooth running of their team. However, such a manager allows office bullies free reign for several reasons. One reason could be to avoid presenting oneself as a disciplinarian, as such an image is a negative one, and thinking it better to offload such tasks to another.

This is in keeping with Machiavelli's philosophy. Machiavelli advised a ruler to find a henchman to perform one's unsavory tasks. Consequently, it was wise for the ruler to place bounds on what a henchman can do, and moreover, to remove the henchman if the conditions should no longer prevail.

Hence, a manager who tolerates office bullies need not be a thoroughly bad one if they can place boundaries on how and when they can act, and even better if they can bring them in line if things get out of hand.

If a manager is in control and ensures the workplace is not too chaotic, such a manager might face a slight reproach from employees for turning a blind eye to office conflicts. However, such reproach—even if put into official channels in appraisals— rarely has impact, as the upper management rarely chastises a manager they have confidence in.

In turn, employees will accept that working in a company is no bed of roses, and to top it, one must deal with unsavory players on a regular basis. This is an example of a manager who avoids being hated or despised, though their behavior is far from ideal.

Sadly, such managers who support office bullies often have

little control over them.

Another case, an unfortunately common one, is when a mediocre manager believes giving office bullies a free hand compensates for their poor managerial skills. How long such a manager can survive is a combination of several factors, with the main one being luck.

Notably, in such cases, employees get frustrated at stalled projects and being harassed and browbeaten, eventually losing interest in their jobs and end up leaving. The office bullies the manager expected to be a pillar of support, create further chaos once their easy targets left the company. The manager eventually loses the confidence of the upper management and is replaced or simply asked to leave.

Before concluding this description of Machiavelli's philosophy on how a ruler should avoid being hated and despised, the question can be asked—*how can this directly be used by an engineer?*

The answer is that a vast number of transgressions can be forgiven if they also avoid being hated or despised. To reinforce what was stated earlier, every engineer is expected to be completely devoted to their job with little regard to their personal lives, or for that matter, their physical and mental health. Any deviation from this ideal invites reproach from management, either in the course of projects, or at the time of appraisals. These reproaches usually amount to nothing and are a part of the modern corporate culture.

Apart from these reproaches, *what would cause an engineer to be hated or despised?*

An engineer would be despised if they were grossly incompetent, badly-behaved with very poor social skills, or assumed an unearned air of superiority or entitlement. An engineer would be hated if they repeatedly indulged in theft, either financial in nature or with respect to intellectual property.

Machiavelli believed—at all costs—a ruler should avoid being rapacious and grabbing the property or women of his nobles or of his citizens, a quality he believed to be worse than being a murderer. In a similar manner, being a thief is an ignominious label, and if an engineer begins to live off theft, it will cause their downfall.

In our modern times, financial theft is difficult to hide and when one embezzles, the chances of getting caught are extremely high. This is even more so when one is a mere engineer and not a part of the management. When one is part of the power structure of a company, a certain degree of protection is offered, as the management might be motivated to cover up its own wrongdoings. An engineer should make every effort to avoid embezzlement and financial crimes. Engineers who commit these crimes have little or no protection.

Other forms of theft include intellectual property. There are a number of occasions when engineers might be tempted to steal intellectual property. An example could be when an engineer is leaving a company and assumes the right to take with them work done for the company. Almost always, when an engineer works for a company, all work done by the engineer for the company is the intellectual property of the company.

In a few cases, an engineer leaving a company to join a competitor might be instructed by the prospective company to indulge in this intellectual property theft. Whatever may the reason, such cases can end badly for the engineer, as theft of intellectual property will result in legal action, and even if instructed or motivated by the prospective company, it will be the engineer receiving the legal action.

Another form of theft that may not invite legal action, but if adopted as modus operandi, could cause an engineer to be hated, for example, is stealing the work of others without giving them proper credit. Almost every engineer who ever worked in a company or even a university will come across those who routinely steal the work of others. Such engineers may seem like rapid climbers, but unless they possess other qualities, their growth turns out to be ephemeral.

If one wants to be a respected engineer in the long-term, the trust and faith of other engineers and non-engineers is necessary. One who continuously steals the work of others will be branded a thief and ostracized.

Like Machiavelli's distinction between being hated and despised, I tried to separate these for an engineer as well. When an engineer is hated, they are an easy target of management or of their colleagues. Being terminated or excluded from projects can be a massive disruption to one's career.

On the other hand, to be despised can cause either management or colleagues to lose their trust and confidence, or avoid working with one, which in the long term can also lead to termination or exclusion. As with Machiavelli's description of how a ruler can end up being despised by his nobles, the reasons why an engineer may end up being despised by their colleagues and management are also largely behavioral.

An engineer's greatest asset should be their knowledge and skills—and the enthusiasm to enhance them. The reason I say *should be* is because in these modern times, there is a growing belief that technical skills take a backseat once one is in the workforce, with soft skills taking the front seat.

In the previous chapters, I wrote at length about how untrue and dangerous this statement is. Engineers who believe this eventually lose their competitive edge and are forced into the management to survive. An engineer who possesses mediocre skills and knowledge and can't motivate themselves to continuously improve will gradually gain the reputation of someone who cannot be relied on.

Whether the manager or colleagues cast this perspective, an engineer branded in this manner can become redundant in the long term, as the company can find a better engineer as a replacement. It is important to stress that skills and knowledge are not a final and immovable state attained after education, but rather a continuous process of accumulation and growth. In my experience in industry and academia, successful engineers were self-learners using various methods to learn and relearn.

An engineer can be despised for behaving badly. The definition of bad behavior is loose. A person is judged by those who work and interact with them. This can lead to conflict. What management views as bad behavior might be acceptable to colleagues and vice versa.

Bad behavior includes being rude, short-tempered, or condescending. Some talented engineers behave badly. It could be that the investment of time and energy needed to excel at a technical domain does not leave much time to develop social skills.

Despite engineers behaving badly, unless the behavior was truly atrocious, skills and talent will ensure that colleagues will wish to be associated, and the management will wish them to continue with their jobs. Ensure the balance between skills and bad behavior is not tilted too much towards the bad habits so colleagues and managers will cut ties.

To avoid delving too much into the behavioral aspects that can influence an engineer's career, let me describe one last characteristic that can cause an engineer to be despised.

Most engineers focus on results and tangible progress, rather than creating a positive impression. However, in most cases, the management tends to be closer tuned to the impression created by an engineer rather than the results—unless the results are so remarkable that they speak for themselves.

This can be a source of great frustration to engineers, and in a later chapter, I will talk about this in greater detail. Examples could be innocuous—such as not being in the office on time, not participating in team meetings, not volunteering for activities and many more.

For an engineer, these are non-issues when managers behave like a bunch of school teachers. Thereupon, these will draw mild reproach and unless the engineer stretches to extremes, will rarely result in the engineer being despised by the management.

Machiavelli stated that there is no one-size-fits-all method to governing. In the same manner, a ruler could not be ideal and possess only virtues. Sadly, vices are the only way to deal with evil people. He presented many historical examples describing how rulers used various strategies to defend their kingdoms, some successful and others failing.

Thus, before choosing a course of action, a ruler should examine the situation subjectively. A clear answer may not always exist. Though some arguments are repetitive, Machiavelli shows depth and diversity in his philosophy. He was not only an active politician, but also a keen analyzer of events.

> *Some princes, so as to hold securely the state, have disarmed their subjects; others have kept their subject towns distracted by factions; others have fostered enmities against themselves; others have laid themselves out to gain over those whom they distrusted in the beginning of their governments; some have built fortresses; some have overthrown and destroyed them. And although one cannot give a final judgment on all of these things unless one possesses the particulars of those states in which a decision has to be made, nevertheless I will speak as comprehensively as the matter of itself will admit.*

In terms of the citizens of any principality being armed, Machiavelli wrote extensively on the need for a citizen army. This was described in a previous chapter.

> *...rather when he has found them disarmed he has always armed them, because, by arming them, those arms become yours, those men who were distrusted become faithful, and those who were faithful are kept so, and your subjects become your adherents.*

To sum up the need for a citizen army, Machiavelli stressed that when faced by an external threat, a ruler can place the greatest faith in his own subjects rather than on mercenaries or the armies of another ruler. Therefore, when a ruler arms his citizens, the weapons held by his citizens are indirectly held by the ruler himself. Bravery shown by the citizens will go down in history as the bravery of the ruler, and the sacrifices made by the citizens will be those of the ruler himself.

This of course, will only be true when the ruler is not hated or despised by his citizens and nobles, which was already discussed in this chapter.

Similarly, company management striving to continuously train and enable its employees will have an effective workforce and will feel little need to look for external help either through consultants or by outsourcing their work elsewhere.

Sadly, most managers find ways to avoid training and purchasing equipment for their engineers, thinking such training will be wasted if the engineers leave.

Per contra, in the same way citizens and nobles will not rebel against a ruler who is not hated nor despised, the employees of a company will rarely leave unless they find their workplace toxic.

> *...because I do not believe that factions can ever be of use; rather it is certain that when the enemy comes upon you in divided cities you are quickly lost, because the weakest party will always assist the outside forces and the other will not be able to resist.*

Machiavelli wrote this specifically in the context of medieval Italy which at the time had many factions whose allegiances changed frequently and were always at each other's throats.

This situation was deliberately fostered by several princes to keep other princes from gaining too much power.

Though such power struggles might maintain the status quo for a while, in the face of a foreign invader, these factions are useless, as described by Machiavelli in several chapters on how the King of France, the Pope and the King of Spain took advantage and conquered Italy freely.

In the same manner, a company management fostering internal factionalism is weak and petty.

Divisive employees will not keep management secure.

...that those men who at the commencement of a princedom have been hostile, if they are of a description to need assistance to support themselves, can always be gained over with the greatest ease, and they will be tightly held to serve the prince with fidelity, inasmuch as they know it to be very necessary for them to cancel by deeds the bad impression which he had formed of them; and thus the prince always extracts more profit from them than from those who, serving him in too much security, may neglect his affairs ... that he must well consider the reasons which induced those to favour him who did so; and if it be not a natural affection towards him, but only discontent with their government, then he will only keep them friendly with great trouble and difficulty, for it will be impossible to satisfy them.

Machiavelli's statement about how a ruler, especially a new one, should find supporters, is extremely subjective, but provides interesting insights. Machiavelli believed citizens and nobles are drawn to a ruler through either love or greed or hatred for another ruler.

When the reason is hatred for another or to be liberated from occupation, the new ruler should be circumspect about placing faith in them. The yearning for change can lead to another yearning, because they may later turn against him.

On the other hand, when citizens or nobles are drawn to a ruler, through either affection or admiration, such support can be of greater value, especially when the ruler earns great esteem. In other cases, a ruler should ensure the citizens and nobles are weakened and need the ruler's protection and benevolence. Only then can they be trusted to devote themselves to the ruler.

A similar discussion was presented in a previous chapter on how a new manager should behave, especially when an engineer has been elevated to management. Some will be genuinely interested in fresh blood in the management, or have respect for the new manager's past experience. From these, the manager can hope for genuine support. As for those who wish to see another odious manager leave, the new manager must stay on guard and ensure that they don't threaten the new manager.

In my past experiences, many new managers flounder because they fail to see this difference in engineers in their team or among peers in the management and could not curb hostilities before they took root.

> It has been a custom with princes, in order to hold their states more securely, to build fortresses that may serve as a bridle and bit to those who might design to work against them, and as a place of refuge from a first attack ... Fortresses, therefore, are useful or not according to circumstances; if they do you good in one way they injure you in another. And this question can be reasoned thus: the prince who has more to fear from the people than from foreigners ought to build fortresses, but he who has more to fear from foreigners than from the people ought to leave them alone.

In the quote above, Machiavelli described how fortresses built by the princes of Italian states, though appearing useful at first, might prove to be their undoing. Due to the fickle nature of alliances during those times, many princes resorted to building and maintaining fortresses within their principalities.

The question arises. *How and against whom are these fortresses to be used?*

If they were intended to protect against attacks by foreign invaders, and especially powerful ones, those fortresses were liabilities instead of assets. If an invader captured them, little could the prince do to defend his kingdom.

On the other hand, with fortresses built to put down rebellions or to guard against other rival factions, they served a purpose. Machiavelli gave a few examples of how besieged princes or princesses were able to shelter in these fortresses until help arrived.

> For this reason the best possible fortress is—not to be hated by the people, because, although you may hold the fortresses, yet they will not save you if the people hate you, for there will never

*be wanting foreigners to assist a people who have taken arms
against you.*

Machiavelli advised that though instruments such as fortresses
might provide temporary relief to a ruler, the true fortress lies in
the support of the citizens and the nobles. With their support, a
ruler need not fear a rebellion from within the kingdom, and a
foreign invader will only attack at great peril, a kingdom that is
united and armed.

On the other hand, a ruler who is hated or despised can only
rule through cruelty and oppression. In the face of a foreign
invader, both citizens and the nobility might turn against him.

Machiavelli gave a few examples of rulers who were
exceptions, such as the example of the Roman emperor Severus,
who, being warlike and cruel, used guile and deceit to hold the
throne. But such examples are rare and those rulers who fit them
were either exceptional or greatly lucky.

Thinking about the modern corporate world, it is difficult to
relate to a fortress. The closest example I can offer are companies
that create departments whose sole purpose is to obstruct and
burden employees. Examples of such departments are Human
Resource departments whose roles and powers were augmented
beyond the basic needs of hiring and severance and given a free
hand to formulate policies to deliberately burden employees with
red tape.

Company management benefits from such a department that
has employees running around in circles, with the employees
deluded into believing that the Human Resource department is
there to serve its grievances.

The Human Resources department appears to serve as a
buffer between the management and the employees. Eventually,
such bureaucracy only throttles productivity, and if allowed to
expand without limits, can render the company unable to
compete with rivals.

Just as with fortresses, the best solution for any company is to have a management that is engaged with the employees. A motivated and enthusiastic workforce is the best weapon any company can have when dealing with adversity.

This chapter was one of the most difficult to write because Machiavelli's writings on these topics were diverse and full of historical narratives. However, in terms of relevance to an engineer, this is one of the most empowering chapters in the book.

Machiavelli believed rulers who achieved great deeds were far from ideal rulers, and this was not a coincidence, but rather a necessity arising from having to deal with a mercurial and cruel world. The hurdles a ruler faces could be overcome not through virtue, but by resorting to nefarious acts which are normally vices.

It is unfortunate that most engineers are fed an impractical fairytale of how they must behave in companies which are so far from the real, the results are almost always disastrous. This mindset is sadly the result of many years of brainwashing and not just engineering education, and therefore, is difficult to erase.

Often, to live a satisfying adult life, one must forget some aspects of what we learned during our childhood. When engineers step into the corporate world, they should remember they are essentially stepping into the jungle.

Carrying childhood learnings into this jungle is like showing up unarmed to a gunfight. Just as Machiavelli said, a ruler must know how to do wrong when necessary; and engineers should learn how to commit violations if they wish to hold their ground against their managements.

In previous chapters, I described at length how engineers stand to gain when they invest time and energy in training and in side-projects, even if at the expense of their companies. In this chapter, I further stress about how it is important for engineers to not completely devote themselves to their companies unless the need arises, as such devotion will rarely ever be compensated, but will burn them out.

This is consistent with Machiavelli's philosophy. Rulers should be liberal only when either they are ascending the throne and have need for friends or are distributing that which is being plundered from foreign lands. To be liberal with their own resources or with those of their subjects will lead to them being hated and eventually cause their downfall.

In previous chapters, I described Machiavelli's beliefs about a ruler's cruelty and wickedness. Machiavelli believed cruelty was often necessary unless in the case of a hereditary ruler. For a ruler, it was better to be feared than to be loved.

Machiavelli also described cases where rulers gained the throne through wickedness where without being wicked, they would not have become rulers. Machiavelli justified both cruelty and wickedness in a ruler and only cautioned that a ruler should not be hated or despised.

One would normally be confused as to how cruelty and wickedness would not cause a ruler to be hated or despised, but Machiavelli believed the reasons for being hated or despised were often very different.

In the same thread, I believe engineers could face reproach for several reasons, but, if they are not hated or despised, these reproaches will just be minor irritants.

Machiavelli believed rulers would be hated when they are rapacious and grab the property of their nobles or subjects. In similar lines, engineers are hated when they indulge in intellectual property theft, either from their companies or from colleagues.

Such instances of theft are never forgiven, and if committed against the company, will invite legal action. If committed against colleagues, being ostracized will result. Engineers who commit such theft against colleagues believe they can make it their modus operandi, but with time, their reputation will spread, and they will find themselves excluded. Indulging in theft is something an engineer should judiciously avoid.

Machiavelli believed rulers are despised when they are fickle, frivolous, mean-spirited, or irresolute. He gave many examples of Roman emperors exhibiting these characteristics who were murdered by their armies.

On the other hand, engineers are despised when they are unable to keep up with technology, possess shallow knowledge or simply behave badly.

Though these characteristics will rarely destroy the careers of engineers, they become a hindrance if allowed to grow unchecked.

In the next chapter, I will use Machiavelli's philosophy to describe in greater detail how engineers can hone their personalities to further consolidate their positions.

Of Human Nature

THIS CHAPTER COMBINES a few chapters of Machiavelli's writings—*Concerning the Way in which Princes Should Keep Faith*, *How a Prince Should Conduct Himself to Gain Renown*, *Concerning the Secretaries of Princes* and *How Flatterers Should be Avoided*. The theme of this chapter is like the previous chapter, with a difference being that the previous chapter focused on behavior and actions, while this chapter focuses on intrinsic human nature.

Some of Machiavelli's writings in this context were considered controversial, and many of the quotes in this chapter are among his most famous.

I left this chapter until the very end, as the applications to the engineering world are obvious. However, it makes for a smooth transition to the conclusion that follows.

Like the previous chapter, Machiavelli wrote these chapters with respect to the feudal nature of medieval Europe, with the ever-changing political scenario making it impossible for any ruler to behave with integrity and still survive. He gave many examples of contemporary rulers, though he had to omit the names of these rulers to avoid offending them. In modern times, we see unsavory characters in government and politics. In democracies, during elections, the people have limited power and can hold them to account.

This contrasts with the corporate world with only the most basic checks and balances and where immoral leaders are commonplace. Many quotes appearing in this chapter are used by industry leaders and entrepreneurs to justify the observation that industry is a jungle.

In contrast, it is unacceptable for these teachings to be used by engineers and employees, as they would be branded as compliance violations. In this chapter, we will examine how Machiavelli's teachings can be used by engineers and employees with minimal risk.

> Every one admits how praiseworthy it is in a prince to keep faith, and to live with integrity and not with craft. Nevertheless, our experience has been that those princes who have done great things have held good faith of little account, and have known how to circumvent the intellect of men by craft, and in the end have overcome those who have relied on their word.

This opening statement is similar to the opening statement in the previous chapter, where Machiavelli stated that the real world is so far from the ideal and if a ruler chooses only to follow the ideal, he would sooner or later be ruined.

When he speaks of *faith*, he implies promises or pledges, or in a more general term: integrity. During those times, it was common for rulers to break their promises to one another, especially those made regarding military allegiances.

In the previous chapter, Machiavelli praised Cesare Borgia (also known as Duke Valentino), who was notorious for breaking promises and murdering his rivals after guaranteeing their safety.

One could say that this is obvious during times of political turmoil, however, an astute ruler should know how to break a promise without appearing blatantly two-faced.

> You must know there are two ways of contesting, the one by the

law, the other by force; the first method is proper to men, the second to beasts; but because the first is frequently not sufficient, it is necessary to have recourse to the second. Therefore, it is necessary for a prince to understand how to avail himself of the beast and the man.

On its own, the statement above is something democracy-lovers dread, as the use of force outside the rule of law leads to chaos. In those medieval times, when republics were rare, and most kingdoms were principalities, the ruler and nobles usually operated outside the realm of the law.

Thus, for them to resort to force might seem like a natural occurrence. As stated by Machiavelli in some chapters, except for the hereditary ruler, it was impossible for a ruler to govern without a certain degree of cruelty.

This cruelty could take the form of imprisoning or murdering enemies or potential rivals, and in many cases, these acts were committed by extra-judicial forces without a formal trial. Even in the modern world, political assassinations occur.

Machiavelli declared such actions as necessary for the safety and security of a kingdom. He believed the chaos following a rebellion or upheaval can be far more devastating.

In the past chapters, while justifying the need for cruelty in a ruler, Machiavelli cautioned against a ruler being cruel out of mere whim and fancy. He presented several examples of rulers, particularly Roman emperors, who, due to their cruelty and numerous murders committed by them were eventually killed by the army.

Machiavelli believed that though a ruler may have to commit unsavory acts, the basis of good governance is always good laws. One might be tempted to swing to either extreme, believing the law alone is supreme or that force alone is supreme, but a ruler needs to use both the law as well as force to govern successfully.

Translating this philosophy to the modern corporate world is tricky, and though practiced routinely, it's one of the few reasons business and corporations are branded as evil. In the same way medieval rulers were rarely chosen by the people, and their cruelty could be termed as selfish, one observes that the corporate elite are the privileged, and their bending or breaking the law are despicable.

> *Therefore a wise lord cannot, nor ought he to, keep faith when such observance may be turned against him, and when the reasons that caused him to pledge it exist no longer. If men were entirely good this precept would not hold, but because they are bad, and will not keep faith with you, you too are not bound to observe it with them.*

Machiavelli's support of a ruler being deceitful was controversial at that time, but if one breaks apart the quote above, it is easy to see its practicality. The feudal nature of medieval Europe made it a fertile breeding ground for deceit and treachery, and thence, for any ruler to assume another would have integrity was foolish.

In an ideal world, violations would be few and far between, but in a chaotic world, violations are the norm and the law is irrelevant. Machiavelli's first statement might make one ponder why a ruler should make a promise in the first place when he might find it necessary to break it later.

Sadly, the feudal nature of medieval Europe made it necessary for a ruler to form alliances and pacts with others, and without such alliances, a kingdom could find itself isolated and vulnerable.

By citing previous chapters, a corporation or a business exists for the purpose of making a profit for the owners. When maximizing profits, company management can undermine laws that ensure fair play, may it be in stifling competition or violating labor laws that protect the welfare of workers and employees.

Inherent human greed can never truly be stifled, and for an employee to expect a company to behave ethically is a fantasy.

An employee submitting to every management expectation

leads to a state where an employee has no personal life and is in pitiful physical and mental health. Here, employees will find the law unfairly pitted against them, as companies will find innovative ways to violate the law. Attempts by employees to do the same will attract punishment.

Unionized workers have a greater say in negotiating terms with company management, but a vast majority of workers are non-unionized and even worse, do not possess regular full-time contracts, thereby leaving them in vulnerable positions.

Therefore, translating Machiavelli's philosophy to an employee requires significant modifications. As already stated in the previous chapters, for an employee to violate the law, they must do so *in spirit* rather than *in letter*.

When speaking of the spirit of employment, one speaks of the ideal symbiotic relationship between an employer and an employee where employees devote themselves completely to the work of the company in exchange for safeguards of the well-being of the employees.

As we know, since the latter is not practiced in reality, the former needs to be compromised. In past chapters, I wrote about ways employees can reduce their commitment to their companies by indulging in beneficial activities without jeopardizing their positions with the company.

In this chapter, I will not repeat those strategies, but instead will focus on how engineers should project themselves within their companies.

> *A prince, therefore, being compelled knowingly to adopt the beast, ought to choose the fox and the lion; because the lion cannot defend himself against snares and the fox cannot defend himself against wolves. Therefore, it is necessary to be a fox to discover the snares and a lion to terrify the wolves.*

The above is the most widely quoted phrase from Machiavelli's writings. Machiavelli describes the need for a ruler to be as ferocious as a lion such that no one dares to grab the throne and wily as a fox to sniff out potential rebellions before they grow to become a threat.

Being ferocious alone is not sufficient, as that would imply ruling by violence, which as Machiavelli stated, succeeded in very few cases when the ruler was an exceptional warrior. In other cases, persistent violence breeds hatred a ruler should avoid.

Deceit alone will not sustain a kingdom. To retain a kingdom, a ruler must defend it from within and without. This is only possible with a strong army. Gaining the respect of the army and keeping them subservient is only possible if the ruler is a capable soldier.

> But it is necessary to know well how to disguise this characteristic, and to be a great pretender and dissembler; and men are so simple, and so subject to present necessities, that he who seeks to deceive will always find someone who will allow himself to be deceived.

Machiavelli wrote the above statement as a response to those who argued that deceit would one day have consequences, and that eventually the ruler would lose the support of the nobles and the subjects. Machiavelli was simply brilliant in exposing the base nature of humans who—despite their fickleness—are also inherently gullible. A vast majority would be content being dumb, driven cattle believing the lies they were fed.

For a deceiver, there will be plenty who are willingly deceived. For most of us, the act of surrendering to another provides a degree of comfort unless it also brings pain and suffering.

To this end, Machiavelli believed a ruler can be a pretender and should not worry about being caught in his or her lies. Few are clever enough to detect his lies.

> For this reason a prince ought to take care that he never lets anything slip from his lips that is not replete with the above-

named five qualities, that he may appear to him who sees and hears him altogether merciful, faithful, humane, upright, and religious. There is nothing more necessary to appear to have than this last quality, inasmuch as men judge generally more by the eye than by the hand, because it belongs to everybody to see you, too few to come in touch with you.

Machiavelli believed a conniving and deceitful ruler should spare no effort in appearing to be the contrary. He must broadcast a public image that is the epitome of virtue. The qualities Machiavelli believed a ruler should exhibit are interesting. He believed a ruler should appear merciful to reassure the people that they will not be subject to unjust cruelty and that no one would be punished without justification.

Faithfulness is necessary when a ruler makes a promise or enters an alliance, as only a semblance of being faithful would make a promise credible. To be humane was like being merciful. A ruler should not unleash a reign of terror without reason.

To be upright, a ruler should appear to be driven by a sense of right and wrong. Judgments should not reek of corruption or wickedness—even if they were so.

Machiavelli gave religion great importance. Faith in the divine held sway over people then and still does. Adherence to the tenets of a religion assures the people that their ruler is just.

For that reason, let a prince have the credit of conquering and holding his state, the means will always be considered honest, and he will be praised by everybody; because the vulgar are always taken by what a thing seems to be and by what comes of it; and in the world there are only the vulgar, for the few find a place there only when the many have no ground to rest on.

This is Machiavelli's final assurance with respect to the true character of a ruler. If a ruler could retain his kingdom, ensure peace and prosperity, repel foreign powers and crush internal disturbances, he will be seen as successful.

Most people cannot think critically, and in the end, will only judge by the final result. A concept is: *the end justifies the means.* Machiavelli calls the common folk vulgar if they are swayed by what they see and are unable to look beneath the surface. He says a ruler should use this vulgarity to his own advantage, as their sheer number offers majesty to the throne and dissuades opponents from trying to remove him.

In the above quotes, I did not present an analogy to corporate life, but expanded on what Machiavelli wrote. This chapter has incredible power. I read and reread this chapter dozens of times. In those chapters, Machiavelli speaks of the human character of the ruler in opposition to the previous chapters where he wrote about the courses of action under various circumstances.

This is true in modern times as well. In the end, everything filters down to the individual. Companies and organizations are comprised of humans. The potential of those who form the company dictates the potential of the company. A successful company is comprised of good leaders and skilled employees—one without the other is of very little use.

Let us now translate Machiavelli's philosophy to the modern corporate world, first to judge the management from the point of view of the employees, and next, to understand how an employee can use Machiavelli's philosophy.

Since a great deal was already written about management practices and how engineers can deal with various strategies used by management, the discussion that follows will omit the survival strategies that an engineer can use, but instead will describe how these strategies can be effectively masked.

Everyone who has worked in industry have come across members of the management, who are incredibly suave and manipulative, speaking so eloquently as to brainwash employees and other members of the management. The larger the organization, the larger number of such managers and leaders.

Faced with this constant indoctrination, employees parrot

propaganda and empty jargon becomes commonplace. At this point, many employees accept corporate slogans as universal truths.

Such managers conform to Machiavelli's advice about how rulers should never let slip from their tongues any word that strays from the ideal. These managers will speak of the noble pursuits of the company, of doing business while upholding the highest standards of ethics, of the need of rigorously abiding by laws, and how the products and services of the company are improving the lives of common folk.

Anyone working in a large organization will spend countless hours in useless meetings where managers talk endlessly about the lofty ideals of the company and exhort their teams to devote themselves completely to the company.

In addition to countless managers, mega companies invest in human resource departments with the sole purpose of concocting a myriad of company policies that serve little purpose but add to the swamp of empty jargon.

The manner corporate propaganda is received by employees varies. In almost every company, there are employees who sing the company song and in time lose all semblance of their own personal identities. Machiavelli speaks of those who wish to be deceived. We place this group of employees in this category.

In large, profitable companies that provide their employees with good remuneration and benefits, this category of employees can form the majority. With stable jobs, most employees live in a state of delusion thinking that the company management truly cares about their wellbeing.

Companies achieving this state did so cleverly by using a carrot and stick policy. With good compensation provided to employees, they can be coerced into working long hours to the point of burnout. By doling out a few additional benefits like sponsored picnics or subsidized recreational activities, many employees are content with their jobs.

Of course, managements that continuously stress an *employee first* policy will usually most often be the first to turn around and indiscriminately fire them when their business goes south.

In recent times, mass layoffs were widespread with many large companies cutting hundreds or even thousands of jobs with little or no notice. Management set aside commitments made to their employees and did what had to be done to cut their losses. The company masked its brutal profit-is-everything motive with convenient lies its employees were deceived into believing.

One might ask, *did the management even have to hide their true nature? After all, would it even matter to the management if the employees knew they were dispensable?* A company that invested in deluding their employees of their true nature can soften the blow to their reputation during a mass layoff. This is evident when one sees employees who are laid off taking to social media and thanking their companies for the opportunity provided to them, rather than complaining—as they rightly should.

Furthermore, by deluding the workforce of its true intentions, astute management—through a combination of perks and punishments—can create an intensely competitive atmosphere to extract the most from their employees. Often, the final prize that an employee will win for coming out ahead in the competition is negligible in comparison to the fact that the management has succeeded in pitting each employee alone against the management, where the management always has the upper hand in all negotiations.

In trying to fulfil the management's expectations, the company can manipulate employees into a state where they are bound to the company, and therefore has successfully disempowered them. These disempowered employees burn out, as they find the prospect of adjusting to another company extremely difficult after the time and energy already invested.

For employees who are completely taken in by management's propaganda, they would either not realize they are being exploited or consider this exploitation as nature's law of "survival of the fittest".

Some become bitter when they do not receive proper reward for their hard work, while others search for better techniques to get ahead in the rat race. However, these attempts at self-improvement are equivalent to a gambling addict attempting to win big. The gambling house always wins. Few can win the corporate rat race.

For employees seeing through management lies, many accept the situation as the inevitable sad state of the modern world. Machiavelli says those who know the truth are few and unable to threaten the ruler if the ruler deceives most people.

With the above discussion on how company managements use Machiavellian principles to push their employees into a corner, let us reverse the discussion on how an employee can use those principles. Referring to previous chapters, employees in a company should seek an independent existence for various reasons—greater employability, greater job satisfaction and immunity from toxic office politics.

It is advisable for an employee pursuing independent side-projects or education to project the image of a model employee and avoid raising management's suspicions.

Like Machiavelli's advice to rulers, employees should appear sincere, diligent, loyal and trustworthy, to be thoroughly enjoying their jobs. Employees who convince their managers of their enthusiasm and love for their jobs and being completely devoted to company projects can successfully pursue independent projects with the least interference.

In the same way a company can abandon its promises to its employees, employees can also care little about their commitments made to the company when such commitments conflict with their wellbeing. While profits are paramount to the management of the company, stable jobs with decent compensation and benefits are paramount to employees.

Employees are expected to dedicate all their energy and time to their jobs, and even more so during the time of the interview when a candidate is expected to show the utmost enthusiasm and pledge that they would rather be found dead at their desk rather than lack in commitment. Such enthusiasm is essential soon after beginning a new position, as first impressions are lasting impressions—if not the last impression. Other times to show the highest dedication are at critical times. These times include being close to a project deadline or before performance appraisals.

Once an employee finds their position with a company secure, it is prudent to cast aside the promise of utmost commitment made to the company, and instead prioritize one's own growth instead. This is in contrast to the company's interests that the engineer does not imagine any existence outside the company, even if it implies indulging in non-technical events within the company that the company claims might build interpersonal skills or assimilating in *company culture.*

This is where the engineer must choose between activities that benefit them versus activities that appear beneficial but in reality, primarily serve the company's interests. Nowadays, there are progressive companies allowing engineers to undertake activities within the company's domain, but still can be extremely beneficial to the engineer by adding to their skills or knowledge.

This is still the rarity and not the norm, and when an engineer finds themselves *being kept busy* by management, one needs to ask whether early promises need to be fulfilled.

When an engineer invests a non-negligible portion of their time and energy into augmenting skills and knowledge either through training or side projects, that is a direct investment in their inherent value as an engineer. This is in contrast to being involved with the company, where value addition is indirect, as the primary beneficiary is the company and not the engineer.

In the event the engineer is unable to find time outside of official working hours, they should devote time during official working hours towards self-improvement. This is a classic case of an employee casting aside their promise made to the company in favor of ones' benefit.

Such self-improvement will be of use not only during

turbulent times when an employee might need to look for another job, but also if an employee finds other lucrative openings within the same company.

Following this adaptation of Machiavelli's philosophy of how a ruler should not fulfil promises when no longer necessary, it is important to remember that promises need to be made, and one must appear genuine while making them. When one must break them, they should not appear entitled or haughty.

Machiavelli believed a ruler should appear virtuous to enthrall his subjects, as the common people are swayed by what the ruler appears to be, while only the few who are close to him are able to observe him minutely.

In the same way, though a company exists to make a profit, and employees work to make a living, both need to appear to have a nobler cause. Thus, it is inadvisable for an employee to make it obvious they intend to do the bare minimum necessary to collect a pay-cheque, even if that is the true intention.

A company should use noble rhetoric not only during normal times by pretending to be a parent to the employees, but also during times of upheaval when it needs to let employees go. Though retrenchment may be unavoidable, the manner in which it is done can both reassure employees who remain as well as ensure that in the future, the company will not struggle to recruit when business returns to normal.

Examples of these are when a company provides a reasonable notice period to the employees who are being let go, as well as letters of reference which the employees can use to look for other jobs.

There were instances where managers who cut jobs, connected with their peers in industry to arrange for interviews to be conducted for the affected employees. In such cases, though there will be a certain degree of anxiety among the employees who remain, the fact that the layoff was not abrupt and brutal will prevent a mass exodus.

Poor examples however are headline news when companies

terminate hundreds of employees with immediate effect. Affected employees were not given a formal exit and were either informed of the decision through a text message or an email, or even worse being escorted out of the office without even a chance to say a final goodbye to their colleagues.

Such callous actions by the management only sow distrust and contempt in the employees who were unaffected, resulting in an overall decrease in productivity. Affected employees have taken to social media to announce their status, which only further maligns the company.

In such cases, one wished management had read Machiavelli's philosophy. Even if their greatest concern was profit, should at least pretend to have concern for the wellbeing of the employees.

The same philosophy should be exercised by employees as well. Though their primary objective in being employed by the company is for their financial wellbeing, they must project an image of being a dedicated engineer working for the pleasure of the challenge and to be part of the team. Despite actions which might be contrary—such as diverting a significant portion of their working hours into activities independent to the company— engineers should pretend to enjoy their work and believe in the nobler goals the company boasts. A positive image can help an engineer weather turbulent times, or at least get a favorable reference to look for another job if their position is no longer available.

When practicing deception, consistency is important. Machiavelli said a ruler should never let an impious word slip from his or her mouth. This consistency ensures the deception will be accepted. For those who wish to deceive, there will be those who wish to be deceived. A management that is consistent and persistent in its propaganda can successfully brainwash its employees and guide them toward overlooking nefarious practices.

This is because most employees just want a steady livelihood and would like to believe that their work is meaningful. Therefore, they are willing to be deceived by the company's pledges. Those aware of the actual reality won't broadcast criticism when the masses are content. If a few express discontent, they will be labeled as troublemakers or conspiracy theorists.

In a similar manner, management is inclined to believe their employees are too simple to see beyond their lies and would readily believe false promises—especially when made with a certain degree of sophistication by human resources. For an employee who wants to project an image of dedication, selflessness and commitment, this is easy. Management is happy to believe their propaganda succeeded.

A few employees find deception to be abhorrent and would fight for what they believe are their rights. However, it is much easier to deceive than to battle, and when an engineer can achieve their personal goals through deception, that is a far easier task than achieving the same through open and honest arguments which can be exhausting and fruitless.

With the above discussion on deception, it is the correct time to touch upon Machiavelli's belief in a ruler having a dual personality—*of a fox to discern the traps and a lion to terrify the wolves*. Therefore, when an engineer perceives the company management acting like a tough parent at times and a conniving cousin otherwise, this is the same dual personality Machiavelli suggested.

However, the timing and application of strategies can determine the wisdom present in the management. Not all managers are wise to use the correct strategies at the correct times, either by nitpicking and micromanaging when a pat on the back was due, or failing to reprimand and correct when poor performance needs to be corrected.

For this reason, it benefits employees to understand manipulative strategies as a defense against management's nefarious tactics.

A wise manager is attuned to what occurs in the team,

whether good or bad. The manager plays the role of the fox and perceives trouble before it festers. Most managers I worked with were unable to do so and woke up only when problems grew to a magnitude when resolution was expensive and at times excruciating.

Some made the mistake of being the lion when they should have been the fox. Instead of feeling the undercurrents and thinking of corrective action, they used force to harangue and bully subordinates. As Machiavelli said, always being a lion is useless. A lion cannot perceive the snares and using unnecessary force reduces its value and effectiveness and is therefore useless when the time for force arrives.

The most effective strategy for a manager to be completely aware of all activities within the team is to have an open and effective channel of communication with the team. In contrast, a manager surrounding themselves with sycophants and mediocre employees while dividing the team into factions, will rarely be fully aware of the working condition of the team.

For an engineer, it is extremely important to gauge how cunning a fox the manager is. Ideally, one might wish for a completely honest manager who exhibits none of the characteristics of a fox. Unfortunately, such managers are either rare, because to become a manager one needs to be manipulative to a certain degree, or impractical, since a manager who has no cunning cannot control their team.

When one imagines a lion, one only sees aggression and strength, but in the context of the corporation, this translates to decisiveness, of making sound decisions after careful deliberation and being steadfast.

Unfortunately, when assuming a position of leadership, this can be seen as a license to harass and bully. Rarely does one see a manager using his or her authority productively. Unfocused aggression can lead to a toxic workplace and the numerous problems described in previous chapters. Here, one should differentiate between normal management and the executives who run the company. For executives to retain their position against a potentially hostile group of stakeholders may require a much higher degree of aggressiveness.

To translate Machiavelli's philosophy for an engineer, I believe it is necessary to choose a different set of animals for comparison. A regular engineer need not be a lion to guard their position. Choose the fox and the horse and assume a dual personality. The fox would still be essential to determine how effective the management of the company is, to understand whether one's manager and colleagues are worthy of one's time and energy. Devoting oneself wholeheartedly to a company with poor management and mediocre colleagues will be a waste one will regret later.

An engineer needs to be cunning as a fox with all the different aspects described in the previous chapters. Understand the quality of the company and behave accordingly.

Choose another animal an engineer should emulate—be a horse instead of the lion. An engineer needs dedication, perseverance and agility instead of aggression. Even if an engineer finds themselves in a company with dubious management and mediocre colleagues, they should find independent projects to remain competitive.

Sadly, giving up on technical excellence because other employees are fatuous is something many engineers do and causes frustration for talented engineers who see their knowledge wasted. As cited in previous chapters, in these modern times, one can always find a fulfilling project independent of the company as an ideal sandbox to hone one's skills. The engineers I came across who were truly satisfied with their professional careers were those who continuously chased technical challenges and took pride in achieving good results.

We arrive at Machiavelli's assurance that as long as a ruler is successful in conquering and retaining control of his kingdom, his methods will rarely be questioned, such is also the reality in modern times. One hardly questions a successful person, and some will even go so far as to justify many of their egregious actions.

If a manager leads a team performing to expectations without a great deal of attrition, they are successful. The fact that those achievements involved lying, duplicity and cunning never comes to the surface. As Machiavelli stated, *the world is full of the vulgar who are taken in by what they see and not what they actually are.*

Upper management will judge a manager by visible factors such as profit and employee attrition and will rarely dig beneath the surface if these factors are satisfactory.

The same can be said for engineers as well.

An engineer who retains their position in a company will be viewed as successful by those who know them. The fact that an engineer may be performing ridiculously menial tasks within the company, while employing an array of deceitful practices to continue with independent side projects, will never ever receive excess scrutiny.

If the assigned work is completed to an acceptable degree of precision, the management will rarely reprimand an employee. Now, if the management deliberately puts an engineer under heightened pressure to deplete their energy, what is needed is a bit of deceit and cunning to be able to prevent minor tardiness from escalating into major reprimands.

As stated before in the previous chapters, engineers who retain their position, while also pursuing satisfying projects that may or may not have any relation to their companies, will serve as role models for younger engineers to enter the domain.

> And the first opinion which one forms of a prince, and of his understanding, is by observing the men he has around him; and when they are capable and faithful he may always be considered wise, because he has known how to recognize the capable and to keep them faithful.

This statement by Machiavelli holds true today as well. To govern effectively, a ruler needs to find good advisors who are both capable and faithful. Every decision should be based on data and wisdom. A ruler must rely on advisors who dedicated their lives to a particular domain of study.

The same goes for management.

A manager is responsible for the deliverables of the team. Not only must the team be efficient and dedicated, but there must also be an effective channel of communication between the manager and the team. Thus, the team is not just the executor, but also the advisor to the manager as the manager must be aware of all ongoing and potential issues through feedback from the team.

Sadly, many managers don't listen to their team and think it is only necessary to issue orders. A manager expecting the team to deliver without ever hearing their feedback will rarely recognize talent and will be quick to assign blame. Most engineers will be frustrated working with them.

An engineer must also judge a manager by those whom they hold close. A manager who has good relations with talented and dedicated employees will value hard work and talent. Managers who surround themselves with the mediocre are insecure and shallow. They will rarely reward skills and talent and may only use talented engineers when the need arises to secure their positions.

Like an effective manager surrounding themselves with capable team members, an engineer should seek good leaders so they can learn and grow. If an engineer finds a capable manager who acknowledges and rewards talent, invest time and energy to support that position. An engineer can learn a great deal under a good manager.

If engineers find themselves working for mediocre managers who have no value for skill and talent, they should find mentors independent of their companies. If mentors cannot be found, one should seek capable colleagues and collaborators, who will fill the position of mentors.

As before, if mentors and colleagues cannot be found within the company, an engineer should look for these independently, either from older acquaintances or through effective networking.

> *Because there are three classes of intellects: one which comprehends by itself; another which appreciates what others comprehended; and a third which neither comprehends by itself nor by the showing of others; the first is the most excellent, the second is good, the third is useless.*

The above statement by Machiavelli is both interesting and apt.

Most assume that a ruler must be exceptional. Though there have been extraordinary rulers such as Alexander the Great, there were many others who were not, but still governed successfully.

This is because they recognized capability in others and were able to retain their services. Thus, Machiavelli states that if a ruler himself is exceptionally wise, he can govern with little assistance, but if he is not, he should recognize wisdom in others.

The ruler who has neither wisdom or the ability to recognize wisdom will lose his kingdom. Machiavelli gave an example of a ruler who would take rash decisions without consulting any of his advisors but did not have the fortitude to stand by his decisions and was forced to retract his decisions and ended up raising greater doubts on his leadership.

A common mistake engineers commit is to expect their managers to be exceptional engineers themselves. Use Machiavelli's philosophy and understand that a manager need not be an expert in all engineering processes a team is responsible for.

It is a very rare case when a brilliant engineer leads a team and shows not just engineering skills but also is a competent manager with good communication and organizational skills. If a manager recognizes talent in a team and utilizes this talent effectively, even a mediocre manager can be successful.

Many engineers are quick to dismiss their managers as incompetent because they were not from ivy league universities, and in many cases, these resulted in lost opportunities, as the engineers could have grown under these managers.

Not only do engineers have unrealistic expectations from their managers but tend to be harsh in assessing themselves. Engineers have been quick to surrender their careers because they did not possess exceptional qualifications.

One needs to understand that to build technology, one needs a community, and beyond one's own skills and talent, the results are dependent on colleagues and coworkers. Therefore, even if one does not possess stellar credentials, recognizing collaborators to help achieve good results is a talent engineers need to cultivate.

Engineers tend to only associate with those whom they know through university or through the workplace, and this limits their choice of potential collaborators. Thence, it is wise to expand one's network and reach beyond those whom one has been compelled to study or work with.

> But to enable a prince to form an opinion of his servant there is one test which never fails; when you see the servant thinking more of his own interests than of yours, and seeking inwardly his own profit in everything, such a man will never make a good servant, nor will you ever be able to trust him; because he who has the state of another in his hands ought never to think of himself, but always of his prince, and never pay any attention to matters in which the prince is not concerned.

Machiavelli's test for judging the faithfulness of a servant is not applicable in these modern times. Rarely will a subordinate devote themselves completely and unconditionally to the affairs of the superior.

A manager with unrealistic expectations of devotion from their team will only gain the loyalty of the mediocre, as it is only the mediocre who are willing to sacrifice their own well-being for the manager.

A manager who can think independently, and who does not need a coterie of followers to massage their ego, is more likely to allow team members to work without undue interference. In modern times, a manager must be willing to allow their team a certain degree of freedom if they achieve the team deliverables with an acceptable level of competence.

On the other hand, no competent engineer should devote themselves completely towards either the manager or the company. The only exceptions would be in unusually progressive companies that rigorously reward merit, or in extremely competitive domains, where mere survival in a company demands a very high degree of dedication.

In all other cases, engineers should invest sufficient time either on their own projects or in continuing education and training programs that make them more employable. As described in the previous chapters, a company will eventually focus on profits and will not hesitate to render an employee redundant if the need arises. Therefore, an engineer should not look upon the company as a protector, but instead devise their own survival strategies.

> On the other hand, to keep his servant honest the prince ought to study him, honouring him, enriching him, doing him kindnesses, sharing with him the honours and cares; and at the same time let him see that he cannot stand alone, so that many honours may not make him desire more, many riches make him wish for more, and that many cares may make him dread chances.

Machiavelli's advice that a ruler should reward and honor a servant is obvious. However, the latter part is also important. The ruler should make it clear that the servant is not alone and must compete with others who serve the ruler. Moreover, the greed of the servant should not be unbounded such that he would sacrifice or overlook the well-being of the ruler.

Lastly, the servant should never escape the scrutiny of the ruler, such that either through carelessness or through malice, he should cause harm to the ruler's affairs. Machiavelli says a ruler should not think of his servants as his children, such that he should reward them endlessly without merit or avoid censure if the need arises.

One expects a similar reward policy within companies but in most cases, either they do not exist or are very meagre. Very few companies perform regular evaluations of compensation and benefits with different domains and accordingly adjust the salaries and benefits of their employees.

In most cases, employees will find their compensation and benefits not growing at all, or at a much slower pace with respect to the general growth of compensation and benefits in the broader industry. It is common to see engineers change jobs every few years such that their salaries and benefits see drastic jumps.

However, changing one's employer also has its drawbacks, particularly if an engineer is working in a company where good work is to some degree appreciated and the workplace is not too toxic. Such is the dilemma faced by many engineers, where they find themselves in a comfort bubble, not ideal, but not bad enough to want to leave.

With the primary objective of a company being to maximize its profits, most companies will either let the salaries and benefits of their employees stagnate or will only increase the salaries of those employees whom they would not want to lose.

This is a never-ending dilemma for managements as well—*should they continuously increase the benefits provided to employees at the expense of their own profit margins, or should they risk losing good workers who choose to leave for greener pastures?*

Most companies only wake up to the issue of stagnating wages only when they find employees shuttling jobs between competitors to jump up their salaries to the extent that an employee might make a full circuit around all the competing companies in a domain to result in a salary that was several times higher than their original salary.

Companies bemoan employees leaving for better salaries while themselves claim the right to do all that is necessary to maximize their profits.

It is well-known that the prime cause for employees to leave a company are toxic workplaces and rampant nepotism. In the absence of these, employees will try to continue in their positions unless they find it difficult to achieve their financial goals. Therefore, the best solution management can pursue to prevent attrition is to detoxify the workplace. Furthermore, company management can provide non-monetary benefits to high-performing employees such as partially flexible or remote working conditions which will be highly appreciated and yet have no impact on the profit margins of the company.

To ensure that employees are not driven to desperation due to stagnating salaries or benefits, the company can either introduce performance-driven increments or other incentives such as stock options.

From the above, it seems trivial for a company management to reward dedicated employees. Unfortunately, few companies do so. The reasons are many as has been described in the previous chapters—mediocre and petty managers, a strong culture of workplace harassment that is difficult to change and a general under-appreciation of talent.

In the absence of a rewards program, employees need to find a way to survive in a company. For many, being constantly on the search for another job is unfeasible due to family constraints or a general dearth of jobs of a particular nature in a given domain.

In such cases, employees need to extract their own rewards from a number of different means as described in the previous chapters.

The best way an engineer can reward themselves is by expanding one's skillset. Acquiring a new skill or expertise in a new domain may solve the problem of insufficient jobs in a domain the engineer currently specializes in. If it is possible to perform this training during regular working hours, the engineer has indirectly created for themselves a reward program that can lead to better opportunities.

In the case of employees who are already highly skilled, independent side-projects are the best way to hone one's skills that are not used within the company, and thereby expand future opportunities. One usually hears the proverb *jack of all trades but master of none* and thinks it is best to specialize in a single domain.

However, the best protection an engineer has against being made redundant and the best weapon when hunting for a new position, is to expand one's knowledge, as when the economy is bleak, those who will be retained or recruited are the most skilled.

> *It is that of flatterers, of whom courts are full, because men are so self-complacent in their own affairs, and in a way so deceived in them, that they are preserved with difficulty from this pest, and if they wish to defend themselves they run the danger of falling into contempt. Because there is no other way of guarding oneself from flatterers except letting men understand that to tell you the truth does not offend you; but when everyone may tell you the truth, respect for you abates.*

The above statement is obvious, but also has depth. After talking about the importance of choosing good, capable and loyal servants, Machiavelli talked about the importance of avoiding flatterers. Flatterers need not be incompetent people who feel the necessity to massage egos to protect their positions but could be nefarious and capable people who remain close to the ruler without the intention to serve.

These could be rapacious people and in times of need turn against the ruler, implying not just that their services are useless, but maintaining them in an advisory position could be dangerous.

There is, however, an interesting case related to those who are not ideal servants that Machiavelli also talks about.

Machiavelli says that for a ruler to avoid this plague of flatterers, he should let it be known that the truth does not offend him. However, if the license to speak freely were granted to all, such a ruler would quickly lose the respect of his advisors who no longer fear him and may even despise or look down on him.

Therefore, to ensure that he is still feared, the license to speak the truth should be granted to a select few the ruler has judged to be wise and who will serve him loyally.

Eventually, a ruler should consult with his trusted advisors, probe them with questions, but his final decision taken after deliberation should be irrevocable. This goes back to previous discussion of how a ruler should choose wise and loyal servants and honor them for their good services.

A ruler who takes the advice of flatterers will rarely make sound decisions unless he or she possesses great intelligence and arrives at decisions independent of what has been advised, which raises the question as to why the ruler needs such advisors to begin with.

A ruler taking the advice of everyone will be confused, will vacillate, and will earn the scorn of his advisors. Machiavelli provided an example of such a ruler. Therefore, once again, the choice of advisors and servants is of great importance to a ruler to avoid falling into the trap of rapacious flatterers. The plague of flatterers and manipulators affects not just the corporate world, but every aspect of modern life. We all face constant manipulation.

In all my years as a student and as an engineer, I saw professors and managers collect around them flatterers and incompetents. The reasons are various as already described and originate from the manager themselves being mediocre and shallow.

Furthermore, a manager who accepts the words of such incompetents as the golden truth is dangerous and will lead the team or company to a dire strait. As with the previous case, an engineer should judge a manager by those whom they are influenced by and how impervious a manager is to poor advice delivered by incompetents.

As already discussed, in such cases, it is best to avoid dedicating one's energies and skills and pay lip service to hold on to one's position.

Like managers, engineers need to be cautious about whom they are influenced by. An engineer will find many people offering advice of all sorts—relatives, seniors, colleagues, professors and more. When considering advice from any person, the engineer should determine both the nature of the advice as well as the nature of the person offering the advice. Not all advice is offered with good intentions, and in many cases, the advisor either has ulterior motives or is simply too ill-informed to offer any meaningful wisdom.

One will find such people hovering around with several different pretexts—offering friendship and love in the case of friends and relatives, or protection and guidance in the case of seniors and professors.

However, in most cases, these people seek benefit from the engineer's competence to use it to their own advantage or to influence it in a manner that might bring them future benefits.

I cannot stress how much damage can be caused to the career of a talented engineer due to the poor advice received from manipulators and lightweights.

Unfortunately, I made many poor decisions due to various manipulative colleagues. Holding them responsible for the mess that later resulted is useless and ineffective. I should not have used their advice. An engineer should assess those who attempt to influence them by several factors, and those who fail, should be ignored.

The best qualification of anyone who attempts to influence is how successful their own background is. Someone with a string of successes and set for greater achievements is likely to at least offer a few words of wisdom, and if those words are aligned to what they have achieved, such advice is worth considering.

On the other hand, someone with a mediocre or dubious background who is unlikely to ever be a high-performer might offer random advice gleaned from various sources or worse, might be attempting to exploit a relationship for their own ulterior motives.

Another important characteristic of an influencer is their analytical ability to understand situations. Those with the ability to take into consideration various factors and arrive at well thought-out decisions are those whom one should keep close. Even if one does not take their advice, such people are valuable for their understanding of situations.

On the other hand, those with unfounded biases who do not use sound logical reasoning but rather arrive at decisions through knee-jerk reactions, should be kept at a distance.

It is important to acknowledge that one will come across a far greater number of the bigoted and ill-informed than those who are wise. Therefore, the process of cleansing oneself of the ill-informed needs to be continuous. If engineers want success, they should regularly purge dubious and nefarious people from their network.

> ...that those men who at the commencement of a princedom have been hostile, if they are of a description to need assistance to support themselves, can always be gained over with the greatest ease, and they will be tightly held to serve the prince with fidelity, inasmuch as they know it to be very necessary for them to cancel by deeds the bad impression which he had formed of them; and thus the prince always extracts more profit from them than from those who, serving him in too much security, may neglect his affairs.

The Machiavelli quote above is an interesting statement. He offers a word of caution with respect to rejecting those who might have been adversaries.

As Machiavelli often stated, a ruler must always be on guard against those who might be a threat to his rule and prevent them from gaining too much power or destroying them.

Therefore, while giving power to someone, a ruler should ensure that the power and privileges bestowed are necessary for the person's survival which will cause him to be loyal to the ruler as if those privileges were to be taken away, his survival would be threatened. This could be very subjective. A ruler should carefully examine such benefactors. On the other hand, blind trust will result in those who take for granted the ruler's support and might either not perform their duties or indulge in their own conspiracies.

I experienced similar situations in companies I worked for, where an initially acrimonious relationship between a manager and an engineer was later not only repaired but resulted in a highly symbiotic relationship. This, however, requires shrewdness on the part of the manager. The manager needs to understand the grievances the engineer might have and selectively solve those which cause the engineer the greatest anguish.

The engineer, aware that the manager is responsible for relieving the pain point, accepts that their success is tied to the manager's success. However, these situations are rare. Often, the initiative is taken by the manager, though another aspect is worth examining in greater detail.

It is basic human nature to become complacent. This fact applies to managers—who take competent engineers for granted—and to engineers who neglect their jobs once their positions are secure. To ensure that engineers remain committed, a smart manager may rejuvenate their teams with fresh talent and try out new engineers with the objective of introducing competition within their teams.

A manager satisfied with their team avoids change and is content with maintaining the status quo runs the risk that sooner or later, the team will stagnate. Managers need to be receptive of their engineer's feedback and look for potential means of strengthening bonds which will increase the productivity of the team.

The same is true for an engineer as well.

Though one's position might be secure and comfortable, one does not know what the future holds. It is possible the manager leading the team might leave the company or move to another role within the company, in which case, the working conditions can drastically change.

Therefore, an engineer should continuously look for talented and capable people to connect with. Even if the present situation makes a collaboration difficult if not impossible, the future could make it possible or necessary.

To decide about forming an alliance with another engineer, find out if there is a pain point that can be overcome through the association. Such bonds can lead to interesting outcomes and provide fruitful future opportunities.

> *Nothing makes a prince so much esteemed as great enterprises and setting a fine example.*

Machiavelli dedicated a chapter to how a prince should gain renown. He believed this was essential. If a ruler wants to be called His or Her Majesty, his or her deeds should be extraordinary enough for the common people and the nobles, such that he or she is held in awe.

Machiavelli gave the examples of rulers who achieved renown through military campaigns. A victory won by a ruler is the best proof of competency. People will feel secure in their reign. A few rulers achieved greatness by being great reformers and bringing in new laws and ordinances.

This follows the previous discussion where he stated that only a few who interact at close quarters with the ruler will know him for what he truly is, while the majority will be taken in by what they see and what comes of it.

The same can be said to be the case with a manager as well.

The best approach for a manager to take control of their team is to score a win with which they gain not only the confidence of the upper management but also the team. This is especially the case with a new manager, who may be unknown to the upper management or the team, and therefore, will need to establish their credentials.

In such cases, the manager may need to work alongside the team on a critical project, with which they bond with the team, and furthermore, the upper management, seeing this positive growth in the team, forms a first good impression of the manager.

A manager with a strong technical background should use their technical skills, and if such skills are absent, should attempt to learn quickly to demonstrate to the team that such capability exists in them.

When an engineer thus sees a manager who during critical times can descend to the level of a mere engineer, such a manager is usually a good one, as they are willing to be responsible for the final deliverables of the team. However, to expect a manager to always work alongside the team is not only impractical, but also a hindrance, as during normal times, this results in micromanagement which only disturbs the team.

Conversely, for an engineer to gain the trust of the manager, they should ensure that at critical times, the engineer performs at their best. Following the previous discussions, humans are eager to be deceived. Most prefer the comfort of a lie to the harshness of the truth. Therefore, whether it is a manager seeking the approval of the team, or for the engineer gaining the confidence of the manager, one need only devise an image of steadfastness, reliability, and dedication at critical times, as these are the times when lasting impressions are made.

> *Again, it much assists a prince to set unusual examples in internal affairs, ..., when he has the opportunity, by any one in civil life doing some extraordinary thing, either good or bad, would take some method of rewarding or punishing him, which would be much spoken about.*

One could say the above advice by Machiavelli is like public relations when a leader makes a show of rewarding good and punishing bad.

However, it is human nature to behave well when it is

rewarded and misbehave when it is not punished. In a similar manner, a manager should also find the appropriate time and the medium to show appreciation for good performance and find ways to correct errors or reprimand bad behavior.

A manager who ignores good performance cannot be assured of it in times of need. If they let bad behavior go unchecked, the workplace becomes toxic with employees behaving in a purely selfish manner.

As an engineer, be on guard against the manager who will only acknowledge their sycophants and ignore the others. As stated before, investing time and energy in such a workplace will leave a talented engineer bitter and disinterested.

> A prince ought also to show himself a patron of ability, and to honour the proficient in every art. At the same time, he should encourage his citizens to practice their callings peaceably, both in commerce and agriculture, and in every other following...

In the above, Machiavelli speaks about the duties of a ruler besides war and statesmanship. For a kingdom to flourish, it is essential that commerce and agriculture thrive and for this reason, every kingdom needs enterprising citizens. These enterprising citizens should be allowed to practice their craft in freedom and peace without fear.

This also follows Machiavelli's advice on how a ruler should refrain from grabbing the property of his citizens, or by appearing to be liberal, to tax his citizens to finance liberal spending.

Machiavelli believed that if their property or their honor are untouched, people are usually content. The best guard a ruler has against internal rebellions is to avoid being hated by the people. Therefore, a ruler establishing laws and ordinances that enable the people to succeed and flourish will be held in high esteem.

Social media is filled with stories about how companies should appreciate their employees and reward good performance, while mentoring those who under-perform and allow them to grow.

Everyone knows this is the ideal.

Reality is far from it for reasons described in the previous chapters. However, as already expressed, a prime reason for employees to leave is a toxic workplace and unbearable harassment. If company management eliminates these conditions and allows employees to work in peace, attrition is reduced.

In addition to eliminating a toxic workplace, if a management provides a flexible atmosphere for engineers to pursue their interests, a great deal of contentment arises.

This is like a ruler allowing his citizens to pursue their callings without fear of their businesses being confiscated. The greatest benefit any company can provide its engineers is letting them pursue their interests without fear.

> *Never let any Government imagine that it can choose perfectly safe courses; rather let it expect to have to take very doubtful ones, because it is found in ordinary affairs that one never seeks to avoid one trouble without running into another; but prudence consists in knowing how to distinguish the character of troubles, and for choice to take the lesser evil.*

The above exhortation from Machiavelli says a ruler, particularly a new one, cannot seek eternal peace and tranquility and avoid making difficult decisions. When one avoids taking up challenges to avoid trouble, one only ends up in another form of trouble.

This is no different from the old proverb that *life is not a bed of roses* where one can hope to lie peacefully for as long as one wishes. Therefore, a ruler must not avoid troubles and challenges, but rather be wise in making decisions to reduce risk or yield greater benefits.

Machiavelli said this in previous chapters, giving the example of the Romans who never ceased to prepare for war.

The result?

The Roman empire flourished for a long time. Other

examples described kingdoms that were ruined because the rulers neglected the art of war.

The same can be said to be true of modern life.

Though most parts of the world do not experience violence like in medieval times, our realms are still fraught with a multitude of uncertainties. Except for the most privileged, an adult can rarely sit for a spell and take a deep breath without facing some uncertainty or the other, may it be loss of employment, illness, bereavement, or many more.

The corporate world is a molten cauldron with constant change and little protection for working people. Any manager or employee assuming life will be smooth and steady is certain to be faced with a rude shock. In particular, the past decade was tumultuous with trade wars, a raging pandemic and now an active war. It appears as if the human race is unable to enjoy peace.

Whether a manager or an employee in these modern times, one needs to constantly improve in order to retain one's position. As expressed previously, companies do not care much for the wellbeing of their employees and are primarily concerned with maximizing their profits.

Thus, the only way managers and employees can avoid becoming redundant is to widen and enhance their skillset such that companies will think twice before either replacing them or closing their positions altogether.

When a company needs to cut jobs, management asks how difficult it would be to recruit a similar employee at a later date. As Machiavelli said, such constant work is difficult and tiresome, but is the only way for employees to turn the tables on companies and retain their positions while fulfilling their ambitions.

> ...that the prince who relies entirely on fortune is lost when it changes. I believe also that he will be successful who directs his actions according to the spirit of the times, and that he whose actions do not accord with the times will not be successful. Because men are seen, in affairs that lead to the end which every

man has before him, namely, glory and riches, to get there by
various methods; one with caution, another with haste; one by
force, another by skill; one by patience, another by its opposite;
and each one succeeds in reaching the goal by a different method.

This is another controversial statement from Machiavelli. He believed a ruler should bend according to the times to retain his kingdom. Tumultuous times may need a bold and adventurous ruler, while in calm and peaceful times, this same rash ruler might lose the support of his nobles were he to seek to upend things others enjoy.

On the other hand, a cautious ruler might be loved by all during times of peace but may fail to quell disturbances during times of turmoil. He gave the example of the Pope Julius the Second, who was rash and impetuous by nature and took advantage of the indecisiveness of all other major powers in the region to attack Italy.

Once he began his military campaign, his allies saw an advantage to assisting him, and with these alliances, his campaign succeeded. Had he waited for a consensus to form, his allies might have found a thousand excuses to deny or delay his plan. However, his rash nature was well-suited to conquering the fractured Italy of those times. Had he lived long and used the same methods following his conquest, he may not have succeeded, as continuous war would have wearied his allies.

One can say the same for modern life. Because changes occur frequently and engineers can find themselves impacted by changes completely external to them, even if they are content with the current state of affairs. Engineers who cannot adapt to the nature of the present time, will either lose their positions, or will not be able to avail of opportunities that are open and within reach.

As an example, engineers who continuously apply for jobs when they are content with their current position are wasting their time. If they leave a non-toxic position, they take an unnecessary risk. Instead, when times are peaceful, engineers should expand their knowledge and hone their skills, either with the official projects of the company, or with independent side-projects. As their profile improves, they are less likely to be terminated when the company is forced to let go of employees or are in a better position to look for other jobs.

This last chapter contains some of the most controversial, but also most quoted passages, from *The Prince*. Machiavelli believed rulers should portray an image of the personifications of virtue, though their actions may be to the contrary.

It is the intrinsic nature of humanity to worship either a divine being or another human as an embodiment of the divine. And in this nature lies our weakness and susceptibility to be deceived. Hence, for rulers to govern successfully, they must enthrall the subjects who are too simple to look beyond what is presented to them.

Engineers who work in industry face many frustrations. They are irritated by the constant barrage of lies from management who claim to be champions of technology, while chasing at nothing more than profit.

They also get upset when management judges them on trivial and shallow aspects such as attitude and openness, while true engineering achievement goes unappreciated. To understand this, engineers should use Machiavelli's philosophy to understand a management, who seek to maintain an image of purity, while at the same time being too shallow to understand the depths of engineering.

As cited earlier, engineers ignore social skills, and thus, can neither understand the management's actions, nor can they act in a manner that satisfies the management.

Sadly, the solution offered to engineers is more training, especially soft-skills training. Though a few of these trainings such as Effective Communication, Project Management, Finance and a few others can be useful, most other non-technical training courses are a feeble attempt to brainwash engineers. Attempting to blend in with corporate culture through these trainings is a complete waste of time and energy. This is a difficult lesson that I learned after working the first few years in industry after completing my graduate studies. It is only after returning to Machiavelli's writings that I truly understood the nature of the game and how an engineer can play it.

For an engineer to continue to be technically competent, and yet, acquire social skills to parley with managers is not only extremely difficult, but actually unnecessary. The engineer needs to understand the true nature of the management through their actions and pay lip service to their lies.

Engineers are not responsible for the operation of the company, and hence, should not concern themselves with issues faced by management. To satisfy the management, engineers should feed the management lies in return. Management wants to be assured of the dedication, commitment and selflessness of engineers, and if engineers can feed them these lies, they would be readily gobbled.

In the same way management chases profits to increase their earnings and bonuses, engineers should attain technical excellence to optimize job security within a company or to have opportunities and good job prospects in the broader industry.

Conclusions

IN THIS CHAPTER, I conclude this book using extracts from Machiavelli's final chapter *An Exhortation to Liberate Italy from the Barbarians*. For those who read this far, I hope it is clear this book is an engineer's survival guide rather than another book on entrepreneurship, as it is in the domain of leadership and management that Machiavelli's philosophy is often applied.

In this chapter, I reiterate why it is important for engineers to find success and contentment—not just for their own personal and professional lives, but also for the betterment of humanity.

> *... then at the present time, in order to discover the virtue of an Italian spirit, it was necessary that Italy should be reduced to the extremity that she is now in, that she should be more enslaved than the Hebrews, more oppressed than the Persians, more scattered than the Athenians; without head, without order, beaten, despoiled, torn, overrun; and to have endured every kind of desolation.*

Machiavelli wrote this book when Italy was fragmented and ruled by rival factions, who in turn had been invaded by first France, and then by the Pope and by Spain.

Thus, Machiavelli made the comparison to the historical enslavement, from whence came great rulers who led them into freedom, such as Moses, Cyrus and Theseus, some of which has been described in the chapter *Starting From Scratch*.

Machiavelli wrote *The Prince* directed to Giuliano de Medici as an exhortation to be this great ruler who would unite Italy and bring them under a single banner, and subsequently seeking a public office in this new kingdom.

His analysis of the historical events that occurred in Italy, the French invasion, and the exploits of Duke Valentino feature greatly in his writings at various chapters in the book.

One might feel the above passage describes events with no relevance to twenty-first century engineering. However, in my experiences in various jobs in industry, the state of most engineers is—if not pitiful—is inadequate and unsatisfying.

Engineers who were my colleagues and friends led unfulfilling and frustrated lives without finding an outlet for their creativity, but instead, performed mundane, trivial and repetitive jobs in toxic environments under managers who have little appreciation.

Some still hope to find that dream job and move from one company to another to find their conditions remaining largely the same, or sometimes getting worse. I wrote extensively about why this state in industry is normal, and how industry leaders feel no need for change. It is the intrinsic nature of business to make a profit, and profits are in most cases made either through mass-manufacturing and economies of scale, or by repetition and automation.

Therefore, *who will liberate the engineers?*

Waiting for an industry leader to create a utopian company where every project will be challenging and every engineer will find an outlet for their creativity, is similar to waiting for a messiah to return.

Each engineer must liberate themselves.

Some engineers believe the only solution to escaping a toxic boss is to become one's own boss and choose the path of self-employment. As stated before, this book is not about entrepreneurship. I am not an entrepreneur. Whether this is the solution to the ennui faced by an engineer is a question I cannot answer.

However, for any engineer deciding to become an entrepreneur, I offer one piece of advice. Entrepreneurship brings a whole new host of issues an engineer working in a company never had to face—looking for customers, advertising products and services, fighting legal battles with competitors and irate customers, and many more. Whether entrepreneurship is the right solution for an engineer is a question that must be answered only by the engineer themself.

In this book, I offer an alternative path which I will summarize below.

One could divide engineering jobs into a few different classifications. At one extreme end are those few engineering jobs in companies operating in an competitive domain with progressive management, and therefore, provide challenging and well-compensated environments for engineers.

Though these jobs exist, they are few—a small number of engineering jobs. Those few engineers who by skill and capability, or by sheer luck, come by such positions, remain in these jobs and lead satisfied and fulfilled lives and feel no need to devise survival strategies.

On the other extreme are jobs from hell where engineers work under abusive managers in companies that expect them to work until they drop dead while providing only the most basic compensation and benefits.

These jobs outnumber the jobs from heaven but are not a majority of engineering jobs. In these positions, there is little engineers can do other than bide their time until another job is available.

Most engineering jobs, however, fall into a middle ground, where they do not provide significant challenges, have a degree of toxicity, and provide average compensation and benefits. Most businesses are based on providing goods and services that do not require a great deal of engineering innovativeness but are still necessary for modern society. Due to this, these companies not only survive, but can thrive.

Since these companies generate significant revenues and profits, they aspire to recruit a skilled workforce because this increases the profile of the company. Engineers working for that company receive reasonably good compensation and benefits but find themselves using few of the skills they spent years acquiring.

On the other hand, that company needs to find a way to keep talented engineers busy. This conundrum was described in detail in earlier chapters.

For the extreme cases, it's easy for an engineer to decide. In a company with challenging work and good pay, the engineer has to work hard and be an asset to the company. For the company from hell, an engineer must find a way to exit as soon as possible by finding another job.

However, for companies in the middle ground, engineers find themselves in a dilemma. To leave a company, they would need a guarantee of a better job, rather than substituting one mediocre company for another.

This brings the rush towards the top few companies with engineers making it their dream to find a way into the mega multi-national corporations. Unfortunately, this situation affects many an engineer who cannot make up their minds to leave, but still find nothing meaningful to do in their current jobs.

On the other hand, companies, especially the larger ones, with several layers of management and needing only a small fraction of engineering skills, have engineers performing an array of menial and mundane tasks. Each manager in the hierarchy in a highly competitive environment will indulge in all forms of office politics to survive.

In several chapters, I described how this situation results in nepotism, factionalism, harassment, and many other evils, all of which are acceptable to the management, who benefit from employees at each other's throats. Many engineers find themselves in companies where they are burdened with tasks, many with little to do with engineering, and are bullied and harassed. Most of us who worked in industry have either faced this harassment or have observed others facing it.

> *...this has happened because the old order of things was not good, and none of us have known how to find a new one. And nothing honours a man more than to establish new laws and new ordinances when he himself was newly risen. Such things when they are well founded and dignified will make him revered and admired, and in Italy there are not wanting opportunities to bring such into use in every form.*

Machiavelli urged the Medici family to create a new order for a unified Italy. In his writings, he referred to contemporary as well as historical rulers, who either created new kingdoms, or who inherited kingdoms in chaos and used valor and ingenuity to establish peace and order.

He wrote extensively about Cesare Borgia (known as Duke Valentino), who he believed was an excellent example of a ruler inheriting a kingdom in chaos but performed admirably in securing it. Besides Duke Valentino, Machiavelli gave other examples of rulers from the distant past, including biblical figures such as Moses, and described the glory to those who forged their own paths.

Many engineers wonder whether it was worth spending countless hours gaining engineering skills, to eventually find themselves using very little of it. Some accept this as a facet of life and resign themselves to being mere salary persons, doing what it takes to survive in the company, while swallowing every form of slight and insult.

I wrote extensively about this in many chapters.

However, if one changes one's perspective and examines the fact that an engineer provides a particular set of services to a company in exchange for salary and benefits and the job does not define everything about the engineer, then as I have described, interesting possibilities emerge. When working in a mundane job, an engineer can invest time and energy into other activities that can either provide opportunities or can lead to side ventures.

In the chapter on *Starting from Scratch*, I described in detail why and how it is important for engineers to take on side-projects. One might think of such side-projects only as commercial ventures, but a successful side-project can be an antidote to problems plaguing engineers in a company.

A side-project can begin as a haven—allowing engineers a sandbox to hone their skills, particularly skills painstakingly gained but without application in their companies. As a side project progresses, it is a wonderful painkiller to office politics and workplace harassment. Engineers immersing themselves in fruitful side-projects will find office toxicity merely irritating instead of causing serious agony. I gave a few examples of side-projects that started from extremely humble beginnings but became mainstream projects.

In this concluding chapter, I once again urge every engineer who finds themselves frustrated in their jobs to not give up on their skills, but rather to embark on side projects that interest them. It is important to remember that not every attempt will bear fruit, and if a few side-projects end up being abandoned, that should not be a showstopper.

Here, it is extremely important to remember that an engineer possesses an advantage that most others do not. They can initiate projects by themselves, possessing the necessary skills to launch a project on their own with limited resources.

On the other hand, an idea originating with a businessperson without a technical background may need greater resources, as the business owner will need to recruit a technical team to realize the idea.

Like the fragmented and occupied Italy of Machiavelli's times, our modern world is plagued with a whole host of problems—climate change, dwindling resources, excessive waste and many more. One could pass these problems on to the government and expect a political consensus to provide solutions.

However, as an engineer, I argue that problems are meant to be solved and if an engineer goes about it in an astute manner, the pursuit of a solution can be empowering and enriching— besides taking humanity a step forward.

Thus, I urge engineers to choose a problem they believe in and work towards a solution. One might think it fantasy, but it is important to remember that as engineers, it is not our duty to solve everything, but the process of arriving at a solution can enrich us—both in technical skills as well as potential opportunities.

For those who argue that engineers find little time outside office hours to indulge in side-projects, I have already described at length the means to cross this obstacle. I described how the mundane nature of most engineering jobs can be taken advantage of, and side-projects can be undertaken alongside company work, rather than execute them strictly outside of office hours.

Though this might seem deceitful and petty, it is important to remember that in these modern times, the boundaries of office hours have been blurred for various reasons, primarily due to the global nature of business.

Therefore, the erstwhile 9-to-5 jobs have dwindled and have been replaced by jobs where one needs to be available late into the night, and also on weekends. Hence, if an engineer multiplexes a side-project with company work, that is not a terrible crime if the official work assigned to the engineer is performed to an acceptable degree of precision.

Most engineers prefer someone to provide them with an opportunity to pursue a challenge. Engineers complain that companies have no interest in taking up challenges and only seek profit. Here, I refer to Machiavelli's quote that *all armed prophets have conquered, and the unarmed ones have been destroyed.*

Therefore, if an engineer seeks the blessing of a visionary leader to launch a project, such an engineer will typically be disappointed as rarely can any business survive without ruthlessly pursuing profits.

On the contrary, an engineer using every possible means to engage in independent pursuits will find true satisfaction; then, success follows from one's own enterprise rather than the blessing of another.

> *With us there is great justice, because that war is just which is necessary, and arms are hallowed when there is no other hope but in them.*

Most engineers find the prospect of committing a violation of company rules to be despicable. Since childhood, we've been taught to do no wrong even if a wrong is done to us.

However, to survive in the modern corporate world, one must deal effectively with those who have few morals and feel entitled to do wrong. If one divides the corporate world into technical people and non-technical people, it's necessary for one party to win and the other to lose.

Usually, it is the non-technical people in management who win because they form the power structure of a company and can write contracts in their favor.

However, when technical people let non-technical people win, that is not a step toward progress. When technical people win and are happy as the result of it, the world is in a better state. A victory for technical people is a victory for innovation and progress. A victory for non-technical people is a victory for greed.

I believe that if any group of people should claim the right to bend the rules, it should be engineers and other technical people, as long as the goals are aligned with innovation and progress.

In the chapters *The Deception of Perfection* and *Of Human Nature* I described in detail—in accordance with Machiavelli's writings—how an engineer can be a fox to detect the nefarious practices of the company management.

To add on, just as Machiavelli stated that a ruler need not fear reproach for being cruel, mean, or deceitful as long as he is not hated or despised, I too believe an engineer should not attempt to be perfect, as long as they are not hated or despised.

In an ideal world, one would never have to do wrong, but in the real world—when one lives among evil—it's necessary to do wrong and know how to do so effectively and only when necessary.

Machiavelli said a ruler who establishes peace in his kingdom and secures it from external threats will be seen as just and honest, for most people are too ill-informed to see beneath the surface and the few who know the truth will not speak out of fear of the majority.

In modern times, we see industry leaders behaving in despicable manners, indulging in egregious actions which—if not illegal—are downright unethical. Yet, the successful are placed on a pedestal and glorified, with some even justifying their methods, writing books, and making movies out of them.

If one wins, no one questions how the victory came to be. Only when one loses do people begin to ask questions. The same can be said for engineers. When one is successful, skilled and can survive in any company, only then do they get praise.

It is only when engineers struggle to hold their jobs that they face scrutiny and receive unsolicited opinions about what they should do to turn their careers around.

> *If, therefore, your illustrious house wishes to follow these remarkable men who have redeemed their country, it is necessary before all things, as a true foundation for every enterprise, to be provided with your own forces, because there can be no more faithful, truer, or better soldiers.*

Machiavelli wrote strongly against the use of mercenaries or soliciting a foreign power to provide arms, and instead believed in a citizen army commanded either by the prince or by another worthy captain.

In the chapter *Technical Weapons,* I described how Machiavelli examined historical events where mercenaries proved either to be cowardly or dangerous. He gave several examples, including that of Duke Valentino, who began by using mercenaries, but after finding them unreliable, resorted to building his own army.

In the chapter *Of Human Nature*, Machiavelli went on to state that when a ruler arms his citizens, those weapons become his, and that citizens who are armed will become faithful to the ruler.

This contrasts with the popular opinion at the time: that arming the citizens could lead to an armed rebellion.

At great length, I criticized the false propaganda fed to engineers about how, once they are in industry, all that matters are soft skills. Companies try to mold engineers to be future managers and through numerous *leadership* programs, try to train them in Communication, Organization and Planning.

Though it would be foolish to brand these soft skills as completely useless, when an engineer accepts soft skills training with little or no time invested in technical training, he or she will become disempowered and unable to survive outside the company. The best way for an engineer to remain competitive and employable is to constantly enhance technical skills and seek out new domains to specialize in.

Machiavelli urged the prince to devote himself completely to the art of war. He gave examples of private citizens who rose to be rulers due to their proficiency in warfare, while some rulers fell because they cared little for the subject of war.

Machiavelli said that no ruler could be safe in the presence of armed soldiers, or command an army, if he himself could not bear arms, and nor could an army be expected to remain obedient to a ruler who himself knew nothing about war.

In the same manner, if engineers wish to make the management think twice about making them redundant, their best approach is to be skilled to the extent that losing that talent would be a blow to the company, and even worse a threat if the engineer joined a competitor. Thus, I urge engineers to engage in continuous technical training and constantly look to learn.

In our modern world, education and knowledge are powerful weapons.

I urge engineers to attain happiness by pursuing supportive friendships and professional connections. As a concluding reminder, when engineers are happy, the world is a better place via innovation and development. Problems and obstacles can be replaced by solutions and hope.

In the chapter *Of Human Nature*, I described in detail how engineers should carefully select their mentors in a manner similar to how a ruler should select his advisors.

This is important. When an engineer selects a mentor, that person should possess intelligence, wisdom, and vision, and should not be rapacious or selfish and cause harm due to selfishness or greed.

It is important for an engineer to ruthlessly weed out those in their network who fail to live up to adequate standards. A few bad apples are all it takes to spoil the basket.

Many engineers hate breaking contact with someone they know, but to this I refer to Machiavelli's quote that *when a ruler is well armed, he will have good allies*. If engineers are skilled and continuously acquire new skills, they will always have people who want to be associated with them if they are not hated or despised.

Many engineers I came across in industry resigned from their professional careers after giving up all hope of achieving success. They quit their profession by just showing up to work and somehow merely surviving in their positions.

Of late, one hears terms such as *lie flat* and *quiet quitting*, where employees gave up and passively decide to take dysfunctional things as they come.

These terms make me sad—because I almost reached that stage. Over the past decade I spent in industry working in several companies, bouncing from one mediocre manager to another, there were times I believed companies were wasting the time of engineers.

However, as Machiavelli urged a ruler to have courage, constantly train and arm themselves, be wise to create new laws and further, be astute to destroy those whose ambitions threaten the throne. Ultimately, I urge engineers to have faith, learn and relearn, and be innovative in launching their own side-projects.

Be a fox to perceive shallow and petty managers and not let them waste your time.

If an engineer stays focused, diligent, invests in themselves and is ruthless in dealing with those who would do harm, I sign off believing success will inevitably follow.